ROBOT ETHIC

JOHN LOK

Made with ♥ on the Notion Press Platform
www.notionpress.com

Contents

Preface

Introduction

This book divides two parts. The first part indicate why (AI) product manufacturers need to concern moral or ethic to manufacture any (AI) products. The second part indicate why scientists need to concern science moral or ethic to do whose scientifical researchs.

In first part, I shall indicate these two questions to concern the difference between (AI) scientist moral and other scientist moral.

(AI) research and development raises two questions about moral responsibilities:

- Should we continue to develop artificial intelligent entities?
- If we do, what are our moral responsibilities to them?

I write this (AI) moral first part aims to let readers to make argument that to judge why robots need have legal, ethic, or moral limitation to concern whether much of robot ethics is concerned with the well-being of human or/and not ethical employment, or moral regard and threaten of robots.

Whether robot ethic or moral or legal questions ought need to be considered by law or society or fully moral agent, and whether any robot ethic or moral or legal issues whether are permitted to enter contracts and largely whether are held fully responsible to (at least some of) their actions.

The most important aim that I expect my readers can make critically logical mind to give yourself opinions to concern whether either which robots will ethical consequence or in this situation, it is not an issue of the morality of the decider, but rather the moral weight of the choice once made.

In AI weapon immoral application aspect, Nowadays, artificial intelligence (AI) technology is popular to be applied to different industry aspects, such as medical, construction, transportation, hospital, education etc. Although, (AI) is a human invention new development. IN fact, it seems only beneficial to human's daily life. But, it will also have threats to influence human's safety in possible , if some scientists or self-interest mind people who aim to apply (AI) to earn more profit or apply (AI) tools to be weapon to attack other countries to achieve to dominate all human's ambitious intention. Thus, (AI) will bring negative influences to our society, instead of positive influences if we can not apply this kind of new technological tools immorally.

I shall give my opinions to indicate what reasons will cause (AI) artificial intelligent tools to be applied to social military defense weapon by human's intention. In my this books, I hope my readers can know what will cause human's immoral behaviors to bring our societies to bring more dangerous or risks or threats if human applied (AI) technology to achieve whose immoral or ambitious intention. Finally, I hope that human ought not apply (AI) technology to do any behavioral attack to satisfy ourselves interest or dominate global world ambition to avoid (AI) technological war occurrence in the future one day. I shall bring the question: Whether AI weapon application is moral or immoral to AI scientist invention.

In my this part, I shall imply some questions are addressed by robots with some moral agency to let my readers to make evaluation of robot its own ethical system and moral judgement.

In second part, what is business science morality? Why is business science morality important? Why do business managers or scientists need to consider whose moral behaviors to achieve to do any decisions as well as consider what are negative or bad influence to other stakeholders?

In my this book second part, I shall indicate attractive cases and examples to explain the reasons why business scientists need to consider their behaviors when they decide to do any matters which relate to their moral mind. I shall explain why scientists need have moral concerning to deal any scientific decision in any business environment suitation. I give ideas to let scientists to know why who need to concern moral and ethic when they are carrying on their science research.

I shall suppose different situations to indicate to these scientists, such as physical, earth, plant, medicine, computer etc. My aim is to raise their judicious ability to judge whether who ought or ought not follow morality to do their scientific research in any situations as well as whether who will encounter what of challenges if who don't follow the morality to do their research in the situation . I hope any scientists can learn how to deal their moral behavior to do correct scientific decision.

This book is suitable to any readers who have interest to pursue to know what bad influences or effects which will be caused to be punished to these business or scientific professionals when their moral judgement are not right to influence the stakeholders' welfares or benefits or hopes seriously. I also let you know what are moral difference between (AI) scientists and other kinds of scientists.

Prologue

Table Of Contents

CHAPTER ONE

Moral safety on artificial intelligence

(AI) scientists have responsibilities to concern moral safety on healthcare, transportation, finance, public policy and defense etc. industries. All safe (AI) systems include relevance of multiple sub- disciplines on these several aspects, such as mechanism design, security, verification, planning, control theory, machine learning, sensor fusion, robust optimization, meta-reasoning, multi-agent systems. All these elements must be any (AI) robot manufacturers need to consider these safety critical systems to ensure any (AI) is used human safety.

Because if any (AI) safety critical systems whose failure could result in loss of life, significant property damage, or damage to the environment. The (AI) safety-critical system failure includes these several aspects. For example, designed or needing to be fail-safe for safety purposes, device or practice that, in the event of a failure responds or results in a way that will cause no harm, or at least minimizes harm; incorporating some feature for automatically counteracting the effect of an anticipated possible source of failure.

(AI) manufacturers need special consideration with (AI) on these several aspects, such as : Due to open-world complexity causes incomplete understanding to bring effects on uncertainties and poor characterization of performance, poor operating regimes, unfamiliar situations; Due to rich ontology of failures causes numerous failure modalities, new attack surfaces (e.g. machine learning attack), self-modification and gaming (e.g. modify reward fun), modeled influences; challenges of transfer across times and space, challenge of coordinating human-machine collaborations, operational capacity. So, any one of these challenges will be any (AI) manufacturers who will face challenges during their (AI) manufacturing process. Also, (AI) scientists or (AI) manufacturers have responsibilities to solve any one of these unpredicted challenges to achieve whose (AI) buyers who can use their any (AI) robot products safely. Hence, (AI) robot safety is (AI) robot scientists or (AI) manufacturers whose moral or ethical responsibilities to their consumers in (AI) robot market.

1.1 (AI) safe test moral reasons

Artificial intelligence (AI) is needed to guarantee a standard level of safety and settle issues, such as compliance with ethical standards and liability for accidents involving, e.g. autonomous (non-manual driving) motor cars. Designing (AI)-based systems for operation in proximity to and/or in collaboration with humans implies that current safety engineering and legal mechanisms need to be revisited to ensure that individuals, and their properties are not harmed and that the desired benefits outweigh the potential unintended consequences.

The different approaches are taken to (AI) safety from pure theoretical (moral philosophy or ethics) to pure practical engineering planes. So, (AI) scientists ought concern that philosophy and theoretical science with applied science and engineering in order to create safe machines. This should become an interdisciplinary approach concerning technical engineering aspects of how to actually create, test deploy, operate safe (AI) based engineering systems as well as strategic, and policy issues. For (AI) smart sensory non-manual driving motor example, non-manual (AI) motor must need to be tested to ensure any person ((AI) non- manual driving car passenger) can sit in (AI) non-manual driving motor safely. It aims to protect any person's life because any person does not drive the non-manual driving motor. So the (AI) engineering must be very safe to help human drivers to drive their non-manual (AI) driving motor on roads.

The best or most reasonable achievement aims to avoid any traffic accident occurrences on roads. Thus, (AI) non-manual driving motor manufacturers have ethic or moral responsibilities to invent the most safe driving machine learning systems modelling and programming, such as deep driving learning systems and programming, reinforcement learning and (AI)-based systems is particularly challenging in safety-critical applications, such as autonomous vehicles, personal care or assistive robots and collaborative industrial robots.

(AI) robot scientists need to intend to explore new ideas on safety engineering for (AI) based systems, ethically design, regulation and standards for (AI) based systems. In particular, they need to spend time to attend any meetings and in depth discussions about different safe (AI) robots issues, bounded morality, safety, safe human-machine interaction and safety considerations in automated decision making systems in a way that makes (AI)-based systems can achieve more trustworthy, accountable and ethically useful functions to any (AI) robot users. Thus, (AI) scientists need to concern diverse communities, such as (AI) safety engineering, ethics, standardization and robotic cyber-physical systems, safety critical systems, and application domain communities, such as (AI) non-human driving automate, (AI) healthcare robots, (AI) manufacturing robots, (AI) agriculture robots, (AI) aerospace robots, critical infrastructures, and (AI) retail robots safe use issues for (AI) users.

However, (AI) safe discussion topics ought concern such as: How to avoid (AI) negative useful effects, safety in (AI) based system design, runtime monitoring and self-adaptation of (AI) safety, safe machine learning, safety constraints and rules in decision making systems, continuous and validation of safety properties, (AI) based system predictability, model-based engineering approaches to (AI) safety, ethically design of (AI) based system, machine-readable representations of ethical principles and rules, (AI) values and goals problem, accountability, responsibility and liability of (AI) based systems, uncertainty in (AI), all safety risk assessment and reduction, loss of values and confidence, self-esteem and the distributional shift problem, reward hacking and training corruption, weapon of (AI) based systems invention avoidance of explanation, self criticism problem, simulation for safe exploration and training problem, human machine interaction safety problem, (AI) applied to safety engineering problem, regulating (AI) based systems, safety standards, (AI) discrimination, human-in-the-hoop and the scalable oversight problem, experiences in (AI) based safety-critical systems include industrial processes, health automate system. All above safe issues will be (AI) scientists who need to discuss how to solve (AI) safe challenges.

1.2 Machine ethics and (AI) robot rights

Machines make ethic decisions or to have rights are misguided. Instead (AI) educators purpose a new science of safety engineering for (AI) agents

(scientists). In particular, (AI) educators need to solve any challenges to the scientific community to develop intelligent systems capable of proving that they are in fact safe, even under self-improvement.

Why is (AI) safe engineering a machine ethics? Nowadays, computer science concerns machine ethics, computer ethics, robot ethics, ethic life, machine morals, and artificial moral issues. Global societies concern with safety of ever more advanced machines, in special concerning intelligent machine safety.

What is (AI) safety engineering? Human-like performance means some immoral actions, which should not be acceptable from the machines (AI) scientists design. (AI) scientists need concern philosophical discussion of ethics for machines be supplemented by scientific work aimed at creating safe machines in a new field we will term " (AI) safety engineering".

(AI) safety research is the possibility of keeping a super-intelligent robotic hardware so as to prevent it from doing any harm to humankind. So, for safety reasons, (AI) systems first be restricted to simulated virtual worlds until their behavioral tendenies could be fully understood under the controlled conditions.

What is safe question to (AI) robots? A safe question is defined as one for which a human being could find an answer without any help from superintelligence, all on hit own merit. Why is this useful? Suppose that a scientist is liking at two potential cures for cancer. Each one could be developed and tested without assistance from the (AI), but would take about one to three years of effort each. Each one could be developed and tested without assistance from the (AI), but it would taken also about one to three years of effort each. If the scientist believes that who has about an equal chance of them. So, if (AI) scientists could ask the (AI)'s assistance. This question is safe for a number of reasons. First, each potential answer has an equal probability of being correct. Second, a human being could answer if without assistance from the (AI) and so asking (AI) for help will only speeding process, but in no ways change the outcome.

CHAPTER TWO

Business ethics of artificial intelligence

What is business ethics of fundamental values and principle for the use of (AI) in business? Its primary goal is to encourage organizations to engage in a multi-stakeholders benefits and considers commitment to values in the application and impact of (AI) developments. What is the impact of (AI) developments on our society? What are the biggest risks that human uses (AI) robots? How will we seek to control the way it affects our daily lives? Are we preparing (AI) robots ourselves sufficiently? What regulations of business ethics to prohibit (AI) robots businesses need to follow?

Nowadays, government more accountability in alike are demanding more accountability in the way (AI) technologies are used, trying to find a solution to the legal and ethical issues that will derive from the growing legal prohibition or legal demanding or legal control or legal limitation of (AI) in people's daily lives.

In business ethics in particular, any (AI) robots manufacturing exporting countries and (AI) robots importing countries both have legal to control (prohibit) the (AI) robots manufacturing organizations' standard manufacturing level or the most reasonable standard of machines; decision making processes or (AI) manufacturing processes and how to ensure that the (AI) systems which adopt always act in a way that is in live with the (AI) robot manufacturing organization's core values. If it is true that with great power, responsibility, (AI) has become increasingly powerful and its applications to business potentially.

However, some (AI) business leaders have indicated that responsible (AI) organizations need to redefine how they interact with technology to be able to be seen as trustworthy in the age of artificial intelligence. (AI) users tend to believe those individuals or institutions that operate with openness to satisfy the public interest. Working with regulators and policy makers, businesses have the opportunity to make a significant contribution to agree on a framework of ethics and norms in which (AI) can be manufactured or innovated safely.

The (AI) business's responsibilities include such as:

Firstly, In the workplace, people often see (AI) not as a tool, they can use to make their job easier or more effective, but rather as something that is done to them and sometimes competes with them to take their jobs. In this case, (AI) company users have responsibility to solve the risk of use from (AI) and to bring the force back on the staffs who are the users of (AI), explaining how company staffs would actually benefit from these technologies and why who shouldn't fear them in any organizations.

Secondly, understanding human values to design values (AI) systems, (AI) technologies are not ethical or unethical per use. The real issues is around the use that business makes of (AI), which should never undermine human ethical values. So, organizations and technology experts have responsibilities to identify the founding values that the ethical framework of (AI) in business.

Thirdly, accuracy issue, companies need to ensure that the (AI) systems to be used to produce, precise and reliable results, no errors. For example, from an unfair sampling of a population, or from an estimation process that does not give accurate results.

The ethical implications of (AI) machine learning are significant. An example is given by the US criminal justice (AI) system, which is increasingly resorting to the use of artificial agents to ease the burden of managing such a large (AI) system. Any systematic bias in these tools would have a high risk of errors, such as (AI) criminal risk assessment system, this software is used in sentencing and hearings across the country. Public argued that the (AI) system misrepresented the risks of different convicts. They suggest that there is a systematic bias on the risk estimation.

So, it is critical that the (AI) machines learning ability that drive (AI) decision making are trained on diverse sets of data in order to prevent similar biases. It is also important that organizations set appropriate accuracy levels to determine clearly their expectations and what an acceptable standard is for them.

Consequently, it is worth noting that in some instances, because (AI) can learn from data gathered from humans. It might be that some human biases are reflected in the machine's decision making.

Fourthly, respect of privacy issue, certainly, machine learning technologies have brought about new ethical issues related to the respect of privacy. Everyone has right to the protection of personal data. Such data must be processed fairly for specified purposes and on the basis consent of the person concerned or some other. Thus, (AI) robot manufacturers need to concern ethics of information.

Opening sourcing material in computer science, when appropriate is an important topic. It helps the development community to under better how (AI) works and therefore be able to explain it more accurately to the public and the media. This is particularly important as better information within the general public improves trust and prevents unjustified fears.

Fifthly, fairness and justice, which are core issues in the stakeholder theory, for (AI) business fairness to all stakeholders and to society as a whole requires (AI) businesses to consider the wider impact of (AI) design and developments. (AI) systems are able to perform tasks, previously undertaken by business in a more efficient and reliable way, the workplace is going to change and it is therefore important that companies pay attention to how this will affect its employers and customers.

Consequently, in (AI) business environment, (AI) has become impact to influence today's business world. In an environment, anywhere new machine learning technologies are created and developed at a fast speed, companies might to be adopted them because they want to be ahead of the game and on top of the latest technological business competition.

CHAPTER THREE

Artificial moral, creation, autonomous and social challenges

What are the reasons (AI) robot manufacturers need to consider artificial morality: autonomy and creativity. I think the main reason why (AI) robot manufacturers need to concern ethic to stakeholders because it is unlike the majority of other implementation of general machine products ethics, if (AI) robots need to follow rules (deontological model) or calculate the best outcome (consequentialism) of an action. It means threat (AI) machine systems need have accurate scientific calculation method and rules to be implemented to satisfy any (AI) users‘ needs. Even, to avoid any accidents to cause (AI) robot consumers' harms.

In fact, (AI) robot can deem as " human behavioral machine only". It means that it is possible humans‘ behaviors can be performed by (AI) machines, e.g. robots algorithm etc. (AI) invention can do complex tasks, e.g. driving cars or school buses or public transportation tools, flying airplanes or drones, fighting on battlefields, recognizing human emotions and expressions. Because (AI) robots can do more complex tasks to replace human and it has more responsibility, so I feel that it has more moral or ethic responsibility to (AI) users.

Are artificial moral agents (robots) possible? I suppose future artificial system can have and do more complex tasks when compared to humans, e.g. future (AI) robots could speak any country's natural language to replace human translator to do translation tasks, e.g. court translation tasks, government translation tasks, language teacher tasks etc. different occupations. Hence, (AI) natural language robots will have more ethic or moral responsibility to let any country people to understand what threat are speaking clearly when (AI) translation robots are doing translation tasks. It is (AI) robot's translation moral responsibility to its school (AI) users when it is used to teach any schools' students or it is used to translate any language to let criminal people to understand in courts. So it is (AI) language knowledge ethic or moral responsibility to (AI) translation robot users.

Hence, it brings these questions? What is a moral action to (AI) robots? What is a moral judgement to (AI) ? Is ethics based on rule-following reasoning to (AI) ? What is moral autonomy in a social context? To answer these questions? One assumption is made that a machine ethics centered on the moral (AI) agent (robot) and its character is preferable to an ethic based on action and moral reasons. However, I think to develop an ethical robot is a challenge of computing.

In fact, autonomy and creative of artificial agents (robots) can be seriously question. Are computers really creators and discovers or are their tools used by humans? Whereas, computerized works of art existed problematic are to assist the process of scientific discovery. What reasons do we have to assume that moral agents (robots) that has enough autonomy, compared to a human agent ? Such as, a soldier in a field needs to make immediate decisions without asking the superiors. So, I assume to view a (AI) robot soldier replaces a human solider in one war. The (AI) robot soldier needs have moral test in the war, whether the (AI) robot soldier is performing better than human soldier in the war. So, in the war case, if one (AI) robot soldier is used to replace human soldier. The (AI) robot soldier has responsible to be used to attack enemy in the war. I believe the (AI) robot soldier has moral or ethic responsibility to protect the country' people life and the (AI) robot soldier have responsibility to attack enemy to achieve goal to fight to win in the war.

3.1 (AI) ethics by design

Ethics by design concerns the methods, tools needed to design (AI) robots with the capability to reason about the ethical aspects of (AI) robots design decisions to guarantee that as (AI) robots' behavior remains within given moral bounds. Can we and should we build ethically aware (AI) robots?

(AI) will have to behave in a way that is beneficial to people beyond reaching functional goals. This is necessary to ensure advanced level of trust between human's and technology, which is needed for (AI) robot users. Hence, ethical, legal and societal issues are raised by the development of artificial intelligence.

(AI) manufacturers need to concern what will be influenced from (AI) robots invention, such as the future of employment, privacy and data protection, surveillance, inter-action with vulnerable people, human dignity, autonomous decision making, the moral responsibility and legal liability of robots between living beings and humans and the status of robots in society.

In all cases, the alternatives being considered can be divided into two types regulation by legation and standards or design ensuring that the (AI) robots themselves take ethical decisions. So, (AI) robot manufacturers need to guarantee different kind of (AI) robot tools, design methods can do ethic (AI) robots' behaviors or performance to satisfy (AI) robot users.

Consequently, can machines make moral decisions? Should artificial systems ever be treated as ethical entities? What are the legal and ethical consequences of human (AI) invention technologies? I think (AI) researchers and practitioners will need to be able to take moral, societal and legal value, concerning how designing of (AI) systems. So, I believe that (AI) robot moral design issue, currently development of (AI) robots has been led by the goal of improving performance and efficiency.

3.2 (AI) Ethical beneficial consideration reasons

To keep societal benefits of (AI) systems, we need to trust them and make sure that they follow the same ethical principles, moral values, professional codes, and social norms that we humans could follow in the same scenario. (AI) is usually defined as the capability of a computer program to perform tasks or reasoning processes that are usually associate to intelligence in a human being.

As an example, playing chess well to need some form of intelligence in a human being, as well as choosing the best in a difficult medical case, or creating something new, such as a mathematical theorem of even some form of art, or even driving a car in the middle of a crowed city.

Why do we need to consider (AI) ethical benefits? Because it depend on what we consider being intelligence in the behavior of a human being at a certain point in time. If our belief about human intelligence changes, and we don't believe any longer that a certain task requires intelligence, then a computer program performing that task is no longer part of (AI), it becomes just another boring computer program. However, (AI) is different general computer program design. It compares the fields of machine learning, reasoning and decision technologies, language, speech and vision recognition and processing technologies, human interface technologies, distributed and high-performance computing, and new computing architectures and devices. When purposefully integrated, these capabilities are designed to solve a wide range of practical problems, boost productivity and foster new discoveries across many industries.

Another reasons why (AI) ethical and beneficial consideration, the presence of (AI) in our current life is popular? For a credit card transaction example, an (AI) algorithm approves that transaction (or not). When we use the GPS in our car, the algorithm that finds the best way to go from where, we are to where be needed to go is called the (AI) algorithm and it is an essential tool for (AI), present or every (AI) teaching book. (AI) robot that cleans the floor of our houses that can work together with humans in production chains, robots that can take tasks of an entire warehouse. Also, self-driving care are all about (AI). They need to be able to see what happens in the street (signals, lanes, other cars, pedestrians, traffic light), they need to able to predict what other cars and pedestrian will do and they need to be able to cope with unforeseen situations. Since, most car accidents are due to human fault. Hence, (AI) system must need have safe ability (effort) to judge how to drive cars on roads. Does it need ethical and beneficial responsibility consideration attentively.

3.3 Law (legal) for (AI) robots

I think that we need rules for such (AI) robots because who can be invented to own human's minds or judgement effort. The legal prohibition of robotic unethical a human being or through inaction allow a human being to be harmed, a robot must obey the orders given if by human beings, expect such orders would conflict to injure human, a robot must protect its own existence as long as such protection does not conflict injure or disobey human. (A Simov, 1984).

I agree above these are ethical laws that human create must advocate the respectful treatment of even those beings / entities that lack moral standing themselves if there is any human's behavior towards other humans might be adversely affected otherwise. Hence, since a goal of (AI) is create entities that can duplicate intelligence human behavior, of not necessarily their form, it is likely that autonomous ethical machines that my be created. If therefore, becomes all the more important that the ethical principles that govern their behavior should be not permit us to treat them badly.

Thus, law (legal) prohibition implementation is controlled to (AI) robots to do behaviors to injure or disobey or attack humans, it is necessary. Hence, human needs to consider how (AI) robots' behaviors will influence our future daily life. The questions include: How do we achieve the aims of autonomous (AI) systems with our own? Does the future of (AI) pose an existential threat to humanity? How do we prevent learning algorithms from acquiring morally objectionable biases? Should autonomous (AI) be used to kill in warfare? How should (AI) systems influence social relations? What sort of ethical rules should (AI) like a self driving car use? Can (AI) systems suffer moral harms? Can (AI) systems be moral agents? How should we live with and understand minds that are alien to our own? All those ethical questions, we need to concern when (AI) robots are used.

3.4 (AI) labor ethical influences

We need to consider how (AI) influences global labor market change. When and in what order should we expect various jobs to become automated? How will this affect the employment and wages of various professions, including less skilled workers, creatives and different kinds of information workers? Because these are (AI) robot manufacturer's social responsibilities to global employment environment. It is possible to influence unemployment number increases of global employers prefer to use (AI) robots to replace human labors (workers).

What has been be historical record on jobs being displayed by automation? What have the average rate and distribution of displacement been, has it been in time, industry and geography? How long before displaced workers found new jobs? Did displacement contribute to inequality? Is there anything different about the advancement of artificial intelligence happening now that would lead us to expect a change jobs being displaced by automation? What factors make an industry to choose automation? Machine historically performed mass-production, but human have been expanding their capabilities with advances in information processing and artificial intelligence. significant parts of the economy, including finance, insurance and many consumer markets could experience how to use (AI) technologies to learn, model, and predict against actions. All above these questions are needed to concern (AI) robot social responsibilities.

CHAPTER FOUR

Does need to keep (AI) legal

Why unassisted human agents, such as auditors, accountants, lawyers, inspectors, police etc. any social occupations which need to be abide by the law. Why does (AI) agents need not to be abide by the law? However, human agents need the assistance of (AI) oversight programs that analyze and oversee operational (AI) programs. So, (AI) agents have responsibilities to be abide by law also. Although, someone argue that (AI) agents also are only machines, they are different from humans who have feeling, judgement and mind effort. But in fact, it is nowadays present, even future one day, (AI) robots are possible invented to own similar to human's mid, judgement ability or analytical thinking effort to do any behaviors or matters by themselves.

For (AI) automatic self-driving car case example, in this situation, whether a (AI) self-driver needs to be abide by law, when this (AI) self machine driver is driving to replace human drivers. Because the car impeded traffic by travelling too slowly. But who could the policeman have cited? The passenger? The car owner? The programmer? The car's computer? In this self-driving car case example, the (AI) computer which had anticipated to drive the car. Hence, it seems that it also needs to be abide by law. In this (AI) self-driving case, it concerns several related questions both in cases of limited harm: Such as, (e.g. a program that causes a driverless car to crash into another and with regard to greater potential harm , (e.g. the fear that smart instruments may rebel against their markets and harm mankind).
So, it seems that (AI) self-driving vehicle has relationship between (AI) and the legal order in traffic criminal law. Because (AI) had anticipated the driving behavior, if it caused any traffic accident occurrence to cause humans or pedestrians to hurt or death or another (other) vehicles to be damaged.

Hence, (AI) manufacturers need to be abide by law to ensure that the (AI) system they use produce covert, precise and reliable results. To do so algorithms need to be free from biases and systematic errors deriving. For example, from an unfair sampling of a population or from an estimation process that does not give accurate results. So, legal can threaten (AI) manufacturers attempt to do any not accurate tests or experiments in (AI) systems manufacturing process. Such (AI) manufacturers data must be processed test for specified purposes and on the basis of the concern of the (AI) manufacturers concerned on some other legitimate basis laid down by law. Hence legal can threaten (AI) manufacturers to manufacture unpassed or fail or not accurate (AI) systems or robot products to sell to (AI) users.

4.1 Ethical and theological reflections on artificial intelligence

(AI) has rapidly grown in different business aspects to be applied. So, in every different (AI) business application aspects, (AI) will bring the ethical or moral consideration to the indicate different (AI) business application aspects. Such as (AI) is applied to eating or food industry aspect (e.g. agricultural planning, distribution, pricing calculation (AI) technological application) ; or (AI) is applied to online relationship communication industry aspect (e.g. face book online communication, mobile apps online-communication); or (AI) is applied to financial technological aspect or (AI) is applied to war (e.g. cyber-defense), even (AI) is applied to be healthcare and education technological tools.

All of these different (AI) industries applications, Which will have unique business ethical Consideration to (AI) robots technology manufacturers. It means every (AI) robots technological manufacturers need to concern what kind of business ethical or moral regulation, who need to prohibit to do what illegal or immoral business behaviors for their (AI) robots invention in order to avoid to be sued by (AI) robot buyers.

Artificial intelligence is not same as artificial consciousness has sometimes been called " strong AI" or " full AI". Because "AI" is different to general Computer products, this products can be invented To own strong or full artificial consciousness feeling or thinking. Hence, this unique consciousness of (AI) high technological products will have more business ethical or moral responsibilities to compare general common computer products, through they are same produced by steels.

As another point, why (AI) has more ethical or moral responsibility to compare general common computer products. Because (AI) systems may have humans for training and/on decision-making ability or effort. Otherwise, computer may not have these unique effort to similar to human's judgement or analytical mind effort. It is one thing for an (AI) to analyze a situation and the make a recommendation to human decision-makers. It is a different thing when (AI) is directly attached to control that allow it to act upon its analyzes without human approval.

As the case, for Amazon publish example, it Applies (AI) technology to help it to publish. Amazon publish recommends a paper book or e-book, (in combination with AI processed data from and other sources) to achieve a target book sale intention for its authors. Amazon does not automatically and autonomously send books to readers and this is a case how (AI) helps it's authors to send e-books or paper books to let readers to buy and read from internet easily. However, (AI) is applied to Amazon publish, due to (AI) helps Amazon publish to make analytical and judgement mind to evaluate how much every book price is the most reasonable price in order to achieve the highest book sale number for every authors, it will follow prior the similar book topic, content and reader reading number sale experience to make evaluation every books' the highest sale price in order to earn the highest sale number.

As this (AI) book publishing case, where bookprice decision making is also automated and is made by (AI) automated calculation system.

Other, (AI) self decision making automated is made by self-driving vehicle case, the entire purpose of a self-driving car is to drive itself, taking the human control driving. This means the (AI) systems must be extraordinary good at decision-making before it can be replaced to human to drive safely. Thus, if (AI) is put in charge of a vital system, like driving a car, and it crashes another car or a pedestrian, than that (AI) might be unsafe. If (AI) is in charge of designed a tall building, and The building falls down. The (AI) might be unsafe.

Hence, such as above three different (AI) application cases, in book publish industry, (AI) manufacturers have moral or ethical responsibilities to design the (AI) systems have the most accurate book price judgement and analytical effort to calculate what the most reasonable or the most attractive reader every book sale price in order to achieve the highest book sale number to help the publish company to earn the highest sale amount. In the non-human driver automate (AI) driving car industry, the (AI) automate driving vehicle manufacturers have moral or ethical responsibilities to design the (AI) driving car which can make the accurate driving judgement effort to drive the car on the roads to avoid any accident occurrence risk to compare to human driving more safe on the road. In the (AI) application building industry, the (AI) building manufacturers have moral or ethical responsibilities to design the (AI) which has the most accurate building design and calculation effort to build the tall building to avoid it falls down easily when external environment factors influences, such as earthquake, serious wind and raining, snow etc. natural environment damages.

Consequently, any different industries of (AI) manufacturers will have different kind of (AI) moral or ethical responsibilities to their (AI) robotproducts. Hence, they must need to concern (AI) ethical or moral issues to avoid (AI) users' complain.

4.2 Social choice ethics in artificial ntelligence

Instead of (AI) manufacturers have moral or Ethical responsibilities to (AI) users, I think (AI) users also have moral or ethical responsibilities to use any (AI) products. This question is that why (AI) consumers who have ethical or moral Responsibilities to use (AI) products.

For (AI) driver example, who must decide how which dangerous collusion to have on the roads. Although, the development of autonomous vehicles puts (AI)s in life-and-death situations, including ethically difficult situation (Lin , 2016). But, (AI) drivers who can't only depend on (AI)'s driving judgement and driving skill effort. They need to know that outcomes of such situations can depend on (AI) design decisions, making (AI) developers to carefully choose the ethics that are built in. Hence, it is the (AI) driver's self responsibility to consider any sudden external environment factors to influence (AI) automatic vehicle's accurate driving judgement effort to cause mislead or careless driving skills.

Implementations of social choice ethics musttake three types of choices, each of which create their one set of ethical dilemmas (Baum) 2009:

- Standing: Who or what is included in the group to have its value factored into the (AI))?
- Measurement: What procedure is used to obtain values from each members of the selected group?
- Aggregation: How are the values of included group members combined to form the aggregated group values?

However, I think that why (AI) users need have social ethical or moral responsibilities as below:
The first reason, with (AI) the social choice process is conducted by machines, not by humans. With (AI) a machine can be sent out into society to
figure out what society wants it to do and then a attempt to do it. Human designers of (AI) must take decisions about standing, measurement and translate this into the technology.

The second reason, there is the question of whether (AI) , itself should have standing arguably, a sufficiently advanced (AI) should. This raises distinct ethical questions between (AI) product Manufacturers and (AI) users. For (AI) autonomous vehicle case , some driving decisions pose tradeoffs between vehicle occupants and other individuals. For example, should the vehicle travel faster, so as to minimize travel time (good for the almost everyone)? One study found a 13% variation is per-mile energy efficiency of autonomous vehicles depending on how the vehicle is programmed (Mersky and Samaras, 2016).

Aggregated across a global fleet of vehicles, this is a serious difference for energy and the environment, a difference that can depend on how the (AI) handles standing. So, autonomousvehicles may be designed to give their occupants the choice of how the vehicle should drive, especially if the occupants own the vehicle. This gives standing to the occupants, but not to everyone when they are sitting in the (AI) autonomous vehicle:

Furthermore, sometimes the vehicle itself will need

to make the choice, because sometimes occupants will neglect to choose themselves, it is their responsibilities when they choose to sit into the (AI) autonomous vehicle.

The (AI) social choice responsibility for (AI) Autonomous driving vehicle case, such as : The vehicle needs default drive settings. An (AI) with social choice ethics would learn from the tendencies of whoever is setting its drive mode and make its own driving choices accordingly.

Designing an (AI) to learn its driving values from its occupants denies standing to everyone. Thus, this could lead the vehicle faster and pollute the environment more, causing the social environment pollution harm.

Hence, it is (AI) autonomous driving car users' responsibilities to give indication to demand the (AI) manufacturers to decide whether who want (AI) help them to drive their vehicle either fast speed (AI) self driving design in order to reduce the time to arrive the destination, but it is unsafe or dangerous to cause accident occurrences and pollute the air or slow speed (AI) self driving design, but it will spend long time to arrive destination, but it is safe to reduce road accident occurrence chances and it will not pollute air easily.

Consequently, it means that (AI) drivers have responsibilities for their life safety and environment pollution when they give indication to the (AI) self driving vehicle manufactures how to design the (AI) autonomous driving vehicles to drive on the roads. Hence, I believe any (AI) users who have social choice responsibilities for their life safety and environment pollution, it depends on how they demand (AI) product manufactures to help them to produce any (AI) products to buy in this (AI) product market.

CHAPTER FIVE

Artificial intelligence and the future of defense

Nowadays, artificial intelligence (AI) is widely knowledge to be one kind of the dramatic technology. However, it is expected to continue, to have a disruptive impact on human's private and public life, so defense and security will be no exception. But how exactly will these be affected ? How will (AI) defense and security is incremental in nature?

To research why artificial intelligence (AI) has possible to be used to cause autonomous weapons by human. We need to understand these three aspects of relationship. They include cybersecurity and artificial intelligence and machine learning and autonomous weapon systems relationship between of them.

Firstly, we need to know what is the mean of artificial intelligence and cyber defense/offense? It means defense of critical networks: real time, pattern finding, anomaly seeking, it must utilize machine (AI) learning algorithms to efficiently, and instantaneously respond to potential network threats as well as it means human on or out of the loop. On the loop : it means anomaly detection: human notified, IT analysis, response. Out of the loop: it means anomaly detection: (AI) decides best method of response: quarantine, honey pot monitoring, hack-back. Thus, it is possible that (AI) can be used , such as autonomous cyber weapon.

What is artificial intelligence and autonomous weapons? Autonomous weapons mean one kind of weapon that can be selected and engaged a target, without intervention by a human operator. Are these machines artificially intelligent? I believe the answer is not, because present weapons systems are not capable of human level reasoning. But, (AI) algorithms are presently employed to process sensor data, monitor system health, take and respond to vocal commands manage data, navigate. This, future autonomous weapons systems will require stronger (AI) to be secure and operationally and cost effective. Moreover, self-aware autonomous cyber systems are crucial.

What is cybersecurity mean? It means the ability to control access to networked systems and the information they contain. It is acted to prevent , detect, recover, react. It is application objects concern people, process, technology and it's application goals are confidentiality, integrity and popular availability. Thus, what is cyber weapon mean? Walware means viruses, Trojans, zero-days, worms ransomware, spyware etc. Does it require a particular objective? E.g. military paramilitary or intelligence. Does it require physical harm? E.g. functional harm or interruption? Mental harm? Is (AI) a technological weapon that it is an object or tool? What about when it is an weapon agent?

In simplicity, (AI) can be one of scientific weapons platform. When one day, it is invented to be applied to control war planes to fly to any countries to attack enemies or it is invented to be seemed to human to replace soldiers to bring guns or any weapons go to other countries to attack. So, it is possible that future any war defense planes, (AI) technological automatic control weapon can be replaced of human soldiers or war plane pilots to control any war defense planes to go to different enemy countries to attack them easily. It is very horror matter to threaten global human's ourselves life in the future , if (AI) automatic control war defense planes or (AI) automatic control machine soldiers were invented successfully.

Hence , when (AI) can be applied to weapons platforms, it structures that launch weapons, i.e. jets, ships, vehicles. (AI) platform and weapon and software architecture components are be done one (AI) technological weapons systems. Thus, human will encounter any (AI) benefits or risks (threats) causes in the same time as soon as possible. If we can predict when (AI) weapon system will be manufactured or invented successfully. Then, we can reduce (AI) weapon systems risks , if we can threaten any (AI) scientists continue to invent any undiscovered (AI) weapons in any time to avoid the future first time (AI) weapon war occurrence in possible.

The (AI) weapon system risk means autonomy: the ability to problem solve technological war , when (AI) weapon system is manufactured successfully, the power to act, how to damage the (AI) weapon system. The power to chance to stop (AI) weapon system manufacturing processes, ability to create a new goals, how to change the (AI) weapon system inventors' or scientists' minds to avoid to apply (AI) tools to achieve attack goals to change to another positive goal. Due to human can't know a prior what an autonomous (AI) weapon system will do.

Although, human is known what (AI) is , but human is also known when (AI) scientists whose emergent behaviors will do to change to do any negative behaviors from positive behaviors. Whatever (AI) weapon system design we use, there will be cybersecurity, problems arising from computation design/complexity. Due to any one (AI) scientist can manipulate the system to act against itself, or who can utilize traditional " cyber weapons" against the (AI) weapon system, or who can manipulate the system to lie to humans, but also due to complexity, there is no way to know if it is lying or not or bounded rationality : satisficing.

Finally, the most serious (AI) technological invention risks are human is unknown these aspects of (AI) absolutely: They are not simple automatic systems, learning reasoning, communication of " self-aware" systems. Thus, human will face (AI) technological invention risks or threats. We need to find any methods to avoid (AI) weapon system is manufactured successfully to avoid (AI) technological war can occur in future anyone day.

(AI) system immoral intention

Why (AI) system can be invented to damage our society ? IS it possible to achieve this (AI) damage system successfully? ON (AI) attribution hand, it can be applied to cars, aircraft, which are subject to regulation designed to protect the public from harm and ensure fairness in economic competition. Thus, (AI) safety issue is important to scientists to consider.

IN general, the approach to regulation of (AI)-enabled products protect public safety issue should be informed by assessment of the aspects of risk that the addition of (AI) way reduce any respects of risk that it may increase. Also, where regulatory responses to the addition of (AI) threaten to increase the cost of compliance, or slow the development or adoption of beneficial innovations, policymakers should consider how those responses could be adjusted to lower costs and barriers to innovation without adversely impacting safety or market fairness.

For example, regulatory challenges that (AI) enabled present are found in the cases of automated vehicles. (AI)s, such as self-driving cars and (AI)-equipped unmanned aircraft systems. IN the long run, self-driving cars will likely save many lives by reducing driver error and increasing personal mobility, it will offer many economic benefits. Thus, public safety must be protected as these technologies are tested and begin to mature. Creating safe spaces and test beds for experimentation , and working with industry and civil society to evolve performance based regulations that will enable more uses as evidence of safe operation accumulates. Thus, it implies that any scientists can also invent (AI) system to control weapon defense planes or (AI) automatic machine human to do any soldier's behaviors to attack to any countries easily, instead of none driver automatic control vehicle invention. Thus, (AI) system can be applied to harm to human or achieve to damage our society aim by ourselves in possible.

The rapid growth of (AI) has dramatically increased the need for people with relevant skills to support and advance the field. AN (AI) –enables would demand a data literate citizenry that is able to read, use, interpret and communicate about data and participate in policy debates about matters affected by (AI). Thus, if (AI) technology is applied to assist human's social development and raising life enjoyment or benefits. It will bring positive impact to influence human's future life. Otherwise, if (AI) technology is unsafe to be applied to threaten human's society. It will bring negative impact to influence human's future life. Thus, (AI) scientists need to consider how to apply (AI) technology.

As (AI) technologies move toward deployment, technical expects, policy analysts and ethicists have raised concerns about unintended, consequences of adoption. Use one (AI) to make consequential decisions about people, often replacing decisions made by human –driven bureaucratic processes, leads to concerns about how to ensure justice, fairness, and accountability, the same concerns of human's safety issue. Thus,)AI) expects have cautioned that there are challenges in trying to understand and predict the behaviors of advanced (AI) systems.

Use of (AI) to control physical-world equipment leads to concerns about safety, especially as systems are exposed to the full complexity of human environment. A major challenge in (AI) safety is building systems that can safety transition from the closed world of the laboratory into the outside open world, when unpredictable things can happen. Adapting to unforeseen situations are difficult necessary for safe operation. Experience in building other types of safety artificial systems and, such as aircraft, power plants, bridges and vehicles has much to teach (AI) practitioners about verification and validation, how to build a safety case for a technology, how to manage risks, and how to communicate with stakeholders about risk. The risk means the harm of human's safety of (AI) damage system control machine invention. Thus, any (AI) scientists need consider moral responsibility when who decide to invent what kind of (AI) system machine to aim to bring human's benefits or attribute to human's welfare intention.

Thus, (AI) products safe invention matter will need any scientists' considerations. Because , if (AI) any products are unsafe or harm human's invention in the manufacturing process, it will bring any human's life danger when the (AI) system damage tools are invented successfully and are provided weapons to humans to use to attack other countries easily. It will cause future global human (AI) technological war occurrence.

I shall recommend the solution is necessary of ethical training for (AI) practitioners and students. Ideally, every student learning (AI) , computer science, or data science would be exposed to curriculum and discussion on related ethics and security topics. However, ethics alone is not sufficient. Ethics can help practitioners understand their responsibilities to all stakeholders, but ethical training should be methods for deciding good intentions into practice by doing the technical work needed to prevent unacceptable or immoral (AI) invention outcomes.

Hence, global human needs to concern (AI) weapon system invention security issue. Nowadays, (AI) has important application is increasing role for both defensive and offensive cyber measures. Currently, designing and operating secure systems requires significant time and attention from experts.

Challenges issues are raised by the potential use of (AI) in weapon systems. The United States has incorporated autonomy in certain weapon systems for decades, allowing for greater precision in the use of weapons and safer, more humane military operations. Nonetheless, direct human control of weapon systems involves some risks and can raise legal and ethical questions concern (AI) manufacturing process intention.

The key to incorporating autonomous and semi-autonomous weapon system into American defense planning is to ensure that U.S. Government entities are always acting in accordance with international humanitarian law, taking appropriate steps to control , to develop standards related to the development and use of such weapon systems. The United States has activity participated in ongoing international discussion on Lethal autonomous weapon systems and anticipates continued robust international discussion of those potential weapons systems. Thus, (AI) scientists have responsibilities to manage the potential to be a major driver of economic growth and social progress only, their (AI) intentions are not the global dominance aims absolutely, if (AI) product industry , civil society, government and the public work together to support (AI) positive development of the technology with thoughtful attention to its potential and to managing its invention threat risks to avoid (AI) products to manufacture to be used weapon tools.

Finally, I recommend that as the technology of (AI) continues to develop, practitioners must ensure that (AI) enables systems are governable, that what their inventions need to be openness to let public to know clearly and understandable; that they can work effectively with people and that their operation will remain consistent with human values and aspirations. Researchers and practitioners have increased their attention to these challenges , and should continue to focus on their future any (AI) inventions.

Hence, (AI) safe system ought to be applied to solve the biggest challenges that society faces, such as mobility for the elderly and those with disabilities, smart buildings may save energy and reduce carbon emissions, precision medicine may extend life and increase quality of life, smarter government may solve citizens more quickly and precisely., better protect those at any immoral invention risk and save money.

Moreover, (AI) enhanced education may help teachers give every child on education that opens doors to a secure and fulfilling life. Thus, these are the future human's potential benefits if the (AI) technology is developed to its benefits and scientists ought avoid to manufacture (AI) tools to cause weapon risks and challenges.

Consequently, the main point is that how experts invent (AI) systems. (AI) systems ought not be advanced weapon systems, it doesn't seem to be thought similar human soldiers mind and behaviors. (AI) system ought be systems that think like humans. (e.g. cognitive architectures and neural networks), systems that act like humans (e.g. pass the test via natural language process, knowledge representation, automated reasoning, and learning), systems that think rationally , e.g. logic solvers, inference and optimization and systems that act rationally e.g. intelligence software agents and embodies robots that achieve goals via perception, planning reasoning, learning , communicating, decision-making and acting function.

In conclusion, it is horror (AI) scientists will invent (AI) systems to be owned human's (soldier's) mind and attack strategic behavior to attack other countries easily, who must need to consider (AI) system ought be invented to own scientists' creating mind and non manual assistance functions for positive attribution to human's society. I expect that (AI) system can only be invented to create human's welfare in our future.

(AI) soldier weapon ethical, social and economic negative impact

In the future, how human can avoid (AI) technological ethical, social and economic negative impact. Scientists need to concern these questions: how to develop of a good (AI) society, how the role and responsibility of the government, the private sector, and the reserch community(including education), in pursuing such a development, whether how the recommendation to support , such a (AI) system development may be in need of improvement.

However, none appers to deliver a comprehensive explicit vision of the role that (AI) system should play in mature information societies. Thus, (AI) 's potential contribution to social good shoud include an in-depth plan for linking in a comprehensive socio-political design questions of responsibility of the different stakeholders, of cooperation between them and of sharable values to understand of a good (AI) positive impact society, not a bad (AI) negative impact society.

Thus, the notion of mature information societies is introduced to stree the importance of addressing the current ethical challenges that (AI) poses in a comprehensive fashion.

It seems (AI) wil invention will be human's moral societal consideration issue. It concerns our (AI) scientists‘ moral issue, how who invent (AI) system to apply to which kind aspects. IF (AI) system was one direction on war weapon tools to similar to soldier's personal mind or attacking behavior. Then, it will bring poor social safety and poor economy growth our world, due to (AI) scientists' moral is low level.

Thus, the developed country US (AI) technological leader needs to focuse on the impacts of (AI)-driven customatin on the US job market and economy. It represents three specific policy responses to the perceived impact of (AI) on the US economy. They include these three aspects such as: How to invest in and develop (AI) for its many benefits, how to educate and train Americans for the jobs of the future and how to aid workers in the transition and empower workers to ensure broadly shared growth.

The future of (AI) influenced cyber conflicts need more than just the application of current and past solutions in order to ensure security and stability of societies, and avoid risks of escalation. To achieve this end, efforts to regulate cyber conflicts require an in-depth understanding of this new phenomenon, identify the changes brought about by cyber conflicts and the information revoluation, and defines a set of shared values that will guide the stakeholders operating to avoid the international (AI) war occurrence. This becomes clear when considering for example, cyber deterrence. Deploying conventional (cold war) strategies to deter (AI)-influenced cyber conflicts proves highly problematic and the urgent need to foster and coordinate new solutions able to account for the any kinds of conflicts of the cyber demain and of mature information societies to avoid (AI) technological war occurrence in the future.

We hope that in the on-going international conversations and reviews, the US government with further specify how " (AI) system invention law" fit into their vision of the future of society in this case the future of (AI) technological war and conflicts. Hence, (AI) scientists need to concern ethical issues related to (AI), like fairness, accountability and social justice can be addressed through increasing needs. Such as: how the creation of a new body focused on robotics and related (AI) system development to avoid to intent to apply weapon tools to provide advice on the policy, legl and consumer protection issues arising in these fields should be considered.

How to achieve ethical training of (AI) staff and ethical education of the public is certainly important responsibility for (AI) tools ethical behavior and design to the private sector and the citizens : of unique challenges that (AI) brings to society in terms in fairness, social equity and accountability are addresses. Thus, the development of the (AI) technology and defining good (AI) remains problematic. In particular, the US government's innovation driven approach to defining the potential, positive impact of (AI) shows that more could be done to ensure that the opportunities and advantages brought about by (AI) are shared by all society.

An initial on Robotics, based upon the ethical framework and guiding principles is proposed. It should be complementary to legislaton and comprise ethical codes of conduct for Robotics researchers and designers, codes for research ethics committees as well as licenses (rights and duties) for designers and users. Thus, (AI) robotics invention of safety issues is very important considertion to any (AI) inventions or researchers. Every country's government ought have legal guiding to control their robotics' manufacturing intention. If their robotics (AI) is applied to seem to be soldiers to attack other countries to threaten their people's safety. Then, those (AI) inventors or researchers need to be punished by law.

In conclusion, I believe (AI) technology will be applied to weapon, when it's technological development is nearly mature to able to learn human's mind to do any behavior. During (AI) technology reachs thie mature stage, I predict the (AI) weapon tool , e.g. (AI) soldiers will have chance to be caused. This (AI) invention mature stage has these characteristics such as:

When (AI) invetion reachs this mature stage, computers and robots will develop conscious, intelligent, personified minds. Further, information technology devices and (AI) systems will be implanted into humans, enhancing, psychological and behavioral abilities and allowing for direct communication with artificial intelligent minds. There will be both artificial intelligence (AI) and intelligence amplification (AI) in the relatively near future stage.

During the (AI) invention reachs this mature stage, these will be an ongoing mulit-faceted integration of information technologies and human life. Humans and information technology will cooperate. Humans will increasingly immerse their lives and minds in (AI) systems of technological intelligence and virtual reality. The distinction between humanity and technology will increasingly close dependence.

During the (AI) invention mature stage reachs that the environment will be infused with information technology, becoming animated, communicative and more intelligent. The destinction between the artificial and the natural will increasing close dependence.

During the (AI) invention mature stage will expand through virtual reality, simulated and virtual reality will increasingly into normal reality, e.g. the (AI) weapons is virtual reality to seem to be soldier weapon.

Finally, during the (AI) invention mature stage is as the global expression of the evolving human-technology integration a " world brain" and " world mind" will emerge on the earth. This psychophysical (AI) weapon system will enhance and enrich the capacities of both individual and collective cogniton. This (AI) weapon system is a potential starting point toward the evolution of a cosmic brain and cosmic mind.

Thus, it is possible that the workship raw data was a unique way in which (AI) could be weaponized to cause war, during the (AI) invention stage reachs the invention mature stage. However, (AI) weapon manufacturing factory will be built possibly. In the future, how will we defins and locate

(AI) weapon factories. Especially, as these factories are no longer solely buildings , but a mil of virtual and substantially different facilities, particularly as it shifts from a physical assemly and development model to a distributed and flexible network. Needing minimal raw materials to develop (AI) weapons, the phsysical location of their (AI) factories could be anywhere and their identification from the outside, nearly impossible. Given the expanding uses for intelligent and super-intelligent (AI). How will we tell the different form a location that is manufacturing (AI) for the creation of weapons versus creating (AI) for an innovative new gaming platform?

In conclusion, human needs to consider every (AI) scientist's personal ethical or moral mind and research intention and (AI) system invention of (AI) weapon factories cause. During (AI) invention reachs the mature stage if human expects to avoid (AI) technological war occurrence in future one day. The technological development on autonomous military robots, ideally among relevant social groups and actors including human-rights, activists, researchers developers, engineers, philosophers, policy-makers, military authorities, lawyers, journalists and the publis need to consider when human has effort to invent autonomous military robots successfully in the future one day. Finally, some ambitious countries or dominant global countries must like to apply (AI) autonomous military robots to be machine soldiers more than human soldiers if (AI) technology had reached the mature stage. So, future (AI) autonomous military robots will be the next choice of weapon to follow nuclear weapon. If civilians were used as a human (AI) soldiers, the weapon simply ignored them and targeted anyway. This scenario highlighted the dangers of proliferation and quick replication of autonomous weapons. Unlike nuclear weapon, a piece of code for (AI) artificial intelligent soldier could be obtained on the black market and replicated at little cost and the hardware for this type of weapon doesn't require costly or hard to obtain components and materials. Thus, (AI) artificial intelligent soldiers can be manufactured many at cheaper cost. Otherwise, manufacturing one nuclear bomb weapon will spend too much cost. Hence , it is possible that (AI) artificial intelligent soldier will be future new technological weapon to follow nuclear bomb weapon. Hence, any country government needs to legislate to control any (AI) scientists' inventions whether they are attributed benefits or welfares to human or damage human's safety.

Consequently, whether (AI) is applied to weapon too. It is moral or immoral intention. It is depends on whether the country intend to use to attack enemy or protect its country. I believe that (AI) is applied to weapon aspect, it is immoral behavior if the country applies it to encourage war to attack the weak economic and social countries in order to achieve the world leader position to control any countries. Otherwise, if the country applied (AI) to be weapon tool to protect itself people's life safety. It is moral invention to the country's (AI) application choice.

CHAPTER SIX

Behavioral ethic in business environment

Macro business society, I believe individuals or team members in organizations who need to consider their activities or behaviors consistently with that they know or believe to be the right thing to do. Because what their intentions will influence their immoral performance to be done whether is viewed legally or ethically. If they do illegal or immoral behaviors and choices, it is possible that they will be punished and they can not predict what and why their behaviors will cause illegal activities by law consequently. For example, an accounting department manager may clearly understand signing a fraudulent accounting statement is legally and ethically wrong, yet who signs the document. Why? What influenced whom to behave unethically? I shall apply scriptive framwork to a social scientific and descriptive mode to analyze what reasons to influence or cause any business professionals or business inventors who choose to do any immoral behaviors neglectly.

In social science view point, it is more important in leading organizations toward more ethical behaviors. My research is related to how ethical decision making any one is made in organizations as well as how to understand ethical or moral behaviors or their behaviors can be better enable managers and leaders to create, and maintain, an ethical culture within an organization.

What is meaning of morality and ethics? When we say a person either acted unethically or immorally, we shall make a distinction between ethics and morality, allowing for a more systematic approach when thinking about ethical issues. In society, all human have responsibilities to do any behaviors which can be responsible to a moral code or some sense of morality. In distinction, ethics means the study of morality from either a philosophical (normative) perspective or a behavioral (social science and descriptive) perspective. Business ethics is the study of morality in the environment of business organizaions, including both normative and behavioral approaches.

How to judge ethic in business science aspect? An assumption that learn moral and norms, such as this nuclear manufacturers and and nuclear science business case. I assume one nuclear scientist intends to invent nuclear energy or power to be pushed space shuttles (rockets) to fly to space, so the the nuclear scientist's intention is to manufacture and improve any new nuclear energy for aim. It must be moral behavior to aim to achieve space exploration missions for human ambition absolutely . Otherwise, if the nuclear scientist intends to invent or improve nuclear bomb to help any ambitious countries' leaders to assist them to attack other countries. It must be immoral behaviors to them to cause nuclear war to threat to human's life safety absolutely.

Hence, it has relationship between business and scientific immoral or moral behavior to be caused from the scientists or businessmen. Also, busines ethic is the study of moral issues in the business environment. Science ethic issue is the study of moral issue in the scientific research environment. What is organizational ethic meaning? Organizational ethic issue is the study of moral behavior in organizational environment. Such as nuclear invention and nuclear bomb or nuclear energy product manufacturins case, if the nuclear scientific research organization and nuclear factory manufacturing organization , whose both organizations have right and fair and reasonable moral mind and regulation to manage or control or prohibit whose nuclear scientists' or nuclear workers' behaviors. Then, the nuclear scientists and nuclear manufacturing workers immoral behaviors will be reduced the chance on occurrence.

Can organizational behavioral ethic regulations prohibit a leader or an organization's top level leader whose behaviors whether really matter in terms of the moral behavior of its employees? Do codes of conduct have an impact in terms of whether employees behave morally and ethically? These questions concern the one whole organization's different departments' staff whole moral behaviors and individual staff of herself or himself individual moral behavior which will influence whose whole organization's ethics or moralities. So, one organizational and behavioral ethics is as a discipline issue. Every department's discipline will influence every organization's whole ethics. If the organization's every individual staff behavior is unreasonable and immoral to influence the stakeholders, e.g. clients are influenced to feel bad image to the organization from anyone of the organization's staffs. Then, it will be caused damage to the organization's image , even it will be reduced the number of clients and it profit to the organization from its any employee's immoral behavior, due to it's clients had lost confidence to buy the organization's products or consum its services. Thus, business ethics will be considered by any organization's leader. Due to any organizational staffs who will represent any organization's image , if one organization had individual staff or some staffs, even all staffs to do immoral or unethic behavior. Then, who will influence the organization's brand to be bad in society in possible. Hence, any organizations need to employ moral staffs to serve their organizions absolutely.

It brings this questions: How to find solutions to avoid the factors affect the behavior of the organization's employees? What strategies will work in terms of producing ethical behavior to the organization's employees? Imagine a leader just given responsibility for the investment banking business of a global financial enterprise. The leader is asked whether ethics is emportant in this new role and how who plans whose leading to the organization to behave ethically. The leader requires ethics issue will be a priority, since confidence in the financial industry has been shaken by recent event. When, pressed for specifics the leade says codes and broad value statements are only clear rules and developing systems for monitoring. The leader says that it is especially importnt to have rules in banking, since the financial industry is prone to unethical behavior, further stating that the size of the banking organization doesn't matter as long as there is a strong leader at the high level. The leader will recruit and promote younger people in the organization, since who believes they will be more ethical and respond better to strong leadership, compared to old employees who have their own ideas and often change ideas and require new leadership method.

In conclusion, moral staffs will be intangible important assets in any organiztions, instead of the organization's leaders. Every organization needs to concern organization ethic issues and every organization leader needs to concern individual ethic issues , also every business founder needs to concern business ethic in nowadays society.

Business ethic case

Factoy manager case:

How can business managers know that a new industrial process will be , or might be harmful to workers?

If the workers' health problem are serious to cause they can not produce any products. Whether ought business managers get their plant build and working as soon as possible?

Who should be responsible for the health problems in workers created by industrial toxins?

What about problems created in people who live nearly or downwind or downriver from an industrial plant?

What are the most dangerous industrial jobs today of these dangerous jobs and their health and safety risk?

Can you think of any way to reduce those risks?

What are the health dangers of other types of jobs today?

Hence, the factory manager's moral responsibility needs to know what the new industrial process caused to harmful to factory workers etc. any reasons which related to influence this factory's harmful to workers before you decide to do this investigation. Because you are one medicine scientist, you must need to know what the factors are to influence the harm to this factory's workers, then you need to decide to manufacture what kinds of medicines to reduce their harm to cause their bad health to their bodies in the future. Due to the money spending is very much, if your judgement is wrong to be recommended to manufacture any unused medicines, you will cause your organization to lose much money unfairly. You need to spend time to gather data about how many prior similar medicines are used to the plant workers can be dealt successfully. Then, you need to find what animals can be used to be tested for your new medicines invention in morality. Hence, it is your moral responsibility to your business managers and yourself.

Drug scientist case:

Supposing you are one scientist to carry on researching how to find one kind of drug to kill a wide of variety of disease in the laboratory. You need to find an antibiotic capable of killing a wide variety of disease producing fungi. But the killer you found had to be gentle enough to not harm the human patients it was to be used on. To find this one miracle drug, you had to test eighty different antibiotics each on thousands of dangerous fungi both in laboratory dishes and in live test animals. Do you feel that it is moral to hurt or kill any animals because you need to use these animals to test this one miracle drug's ability whether it can kill a wide variety of disease producing fungi.

Environment scientist case:

Supposing one environment scientist to research river ecosystem on a river. You need to examine over 500river sections world -wide and you need to lead in the effort to use modern technology to protect rivers and minimize waste dumping to preserve and protect America's freshwater resources. Hence, you need to deal with river ecology and river pollution and research river ecology and see how much variation, you can find in the structure of healthy river ecosystem. You need to concern these issues, such as:

What are the major elements in a healthy river ecosystem?
Should those major parts or niches be in every river and stream?
What are the major resources of river pollution?
What is the difference between point and non point sources of pollution?
When do pollutants do when they enter a river?
How do they harm ecosystem?

Because the environment scientist needs have responsibility to know what the major elements are in a healthy river, what the major resources of river pollution etc . these questions. Although, your aim is to protect rivers and minimize waste dumping to preserve and protect America's freshwater resources for this investigation. But, due to your background is environment science studying, you need to get confidence and fund to assist your investigation from your organization and public support for long time.

Thus, he/she has moral responsibility to prepare this investigation. Due to, who needs to deal with river ecology and river pollution and research river ecology and see how much variation, you can find in the structure of healthy river ecosystem in this New York river location. Hence, you need to gather data to record about where the 900 river sections of different locations in this New York river location from map. Then, you can spend less time to find these 900 river sections in this New York river more easy.

Supposing one environment scientist needs to carry on researching in the desert. How can a scientist document the impact of human noise pollution on a desert species like a desert ? he/she needs to do it by craw through the night desert with a microphone and a set of earphones, trying to avoid rattlesnakes, scorpions, and the shape and cactus needles etc. poison of animals. Hence , you are very dangerous to enter the desert to carry on researching. Your results hoping to help change state and federal development policies in the deserts and protect greater land areas from noise encroachment. Your work with marine mammals continues up and down the pacific coast and has been equally valuable to state and local government officials. You need to concern these issues before you decide to enter this desert to carry on researching, such as:

What is the difference between a tortoise and a turtle? Research desert tortoises, what do scientists learn of value this way?

Can you find any information in the library or on the internet about bioacoustics?

What is noise pollution?

When and how is noise harmful?

Does noise pollution ever affect you?

Make a list of all the unpleasant and harmful noises you hear in a day.

Can you imagine what affect those noises night have on an animal that could not understand where they came from?

Hence, the environment scientist will face danger when who lives in desert. Due to he/she hopes to help change state and federal development policies in the deserts and protect greater land areas from noise encroachment.

He/she must have confidence to find the method(s) to protect greater land areas from noise encroachment in desert, then giving recommendation(s) to this country's government to solve this challenge. Hence, your finding must be used to assist this country's government to solve this challenge for long time successfully.

He/she needs to find the different locations at this desert in order to reduce time to spend to find any valuable places to carry on researching more easy . He/she needs to spend time to stay in the non valuable places. Beside, he/she also needs to own ecology knowledge to prepare this research. For example, he/she needs to learn this ecology knowledge and concept before you decide to do these environmental investigation. Such as, ecology is the scientific analysis and study of interactions among organisms and their environment.

It is an interdisciplinary field that includes biology, geography, and Earth science. Ecology includes the study of interactions organisms have with each other, other organisms, and with abiotic components of their environment. Topics of interest to ecologists include the diversity, distribution, amount (biomass), and number (population) of particular organisms, as well as cooperation and competition between organisms, both within and among ecosystems. Ecosystems are composed of dynamically interacting parts including organisms, the communities they make up, and the non-living components of their environment. Ecosystem processes, such as primary production, pedogenesis, nutrient cycling, and various niche construction activities, regulate the flux of energy and matter through an environment. These processes are sustained by organisms with specific life history traits, and the variety of organisms is called biodiversity. Biodiversity, which refers to the varieties of species, genes, and ecosystems, enhances certain ecosystem services. Ecology is not synonymous with environment, environmentalism, natural history, or environmental science. It is closely related to evolutionary biology, genetics, and ethology. An important focus for ecologists is to improve the understanding of how biodiversity affects ecological function.

Hence, as who is one environment scientist, who have moral responsibility to knw how to do this environment scientific research, such as below:

Life processes, interactions, and adaptations, the movement of materials and energy through living communities,the successional development of ecosystems ,the abundance and distribution of organisms and biodiversity in the context of the environment.

Also, due to ecology is a human science as well. There are many practical applications of ecology in conservation biology, wetland management, natural resource management (agroecology, agriculture, forestry, agroforestry, fisheries), city planning (urban ecology), community health, economics, basic and applied science, and human social interaction (human ecology). For example, the Circles of Sustainability approach treats ecology as more than the environment 'out there'. It is not treated as separate from humans.

Organisms (including humans) and resources compose ecosystems which, in turn, maintain biophysical feedback mechanisms that moderate processes acting on living (biotic) and non-living (abiotic) components of the planet. Ecosystems sustain life-supporting functions and produce natural capital like biomass production (food, fuel, fiber, and medicine), the regulation of climate, global biogeochemical cycles, water filtration, soil formation, erosion control, flood protection, and many other natural features of scientific, historical, economic, or intrinsic value. Besides, the scope of ecology contains a wide array of interacting levels of organization spanning micro-level (e.g., cells) to a planetary scale (e.g., biosphere) phenomena. Ecosystems, for example, contain abiotic resources and interacting life forms (i.e., individual organisms that aggregate into populations which aggregate into distinct ecological communities).

Ecosystems are dynamic, they do not always follow a linear successional path, but they are always changing, sometimes rapidly and sometimes so slowly that it can take thousands of years for ecological processes to bring about certain successional stages of a forest. An ecosystem's area can vary greatly, from tiny to vast. For example, a single tree is of little consequence to the classification of a forest ecosystem, but critically relevant to organisms living in and on it. Several generations of an aphid population can exist over the lifespan of a single leaf. Each of those aphids, in turn, support diverse bacterial communities.

The nature of connections in ecological communities cannot be explained by knowing the details of each species in isolation, because the emergent pattern is neither revealed nor predicted until the ecosystem is studied as an integrated whole. Some ecological principles, however, do exhibit collective properties where the sum of the components explain the properties of the whole, such as birth rates of a population being equal to the sum of individual births over a designated time frame.

The environment scientist also have moral responsibility to own climate engineering knowledge to do this environment scientifical research, such as:

Commonly referred to as geoengineering, also known as climate intervention is the deliberate and large-scale intervention in the Earth's climatic system with the aim of limiting adverse climate change. Climate engineering is an umbrella term for two types of measures: carbon dioxide removal and solar radiation management. Carbon dioxide removal addresses the cause of climate change by removing one of the greenhouse gases (carbon dioxide) from the atmosphere. Solar radiation management attempts to offset effects of greenhouse gases by causing the Earth to absorb less solar radiation.

Climate engineering approaches are sometimes viewed as additional potential options for limiting climate change, alongside mitigation and adaptation. There is substantial agreement among scientists that climate engineering cannot substitute for climate change mitigation. Some approaches might be used as accompanying measures to sharp cuts in greenhouse gas emissions. Given that all types of measures for addressing climate change have economic, political, or physical limitations as some climate engineering approaches might eventually be used as part of an ensemble of measures.

Research on costs, benefits, and various types of risks of most climate engineering approaches is at an early stage and their understanding needs to improve to judge their adequacy and feasibility. No outdoor solar radiation management projects have taken place to date. Almost all research into solar radiation management has consisted of computer modelling or laboratory tests, and an attempt to move to outdoor experimentation was controversial. Some carbon dioxide removal practices, such as planting of trees and bio-energy with carbon capture and storage projects, are underway.

Their scalability to effectively affect global climate is however debated. Ocean iron fertilization has been given small-scale research trials, sparking substantial controversy. Most experts and major reports advise against relying on climate engineering techniques as a simple solution to climate change, in part due to the large uncertainties over effectiveness and side effects. However, most experts also argue that the risks of such interventions must be seen in the context of risks of dangerous climate change. Interventions at large scale may run a greater risk of disrupting natural systems resulting in a dilemma that those approaches that could prove highly (cost-) effective in addressing extreme climate risk, might themselves cause substantial risk. Some have suggested that the concept of engineering the climate presents a so-called "moral hazard" because it could reduce political and public pressure for emissions reduction, which could exacerbate overall climate risks; others assert that the threat of climate engineering could spur emissions cuts.
Hence, the environment scientist needs to own climate engineering and ecology knowledge before he/she hopes to do any environment and climate

scientific investigation. This is whose moral responsibility to decide to do this investigation.

Earth environmentand human behavioural scientist case:

Supposing you are one earth scientist to carry on volcano researching. Imagine hiking down sheer cliff walls into a seething volcano crater filled with hissing fumes and boiling plumes of rising gas and smoke in order to collect needed scientific data. You need to wake such a trip less than three weeks after a small explosion of gas and car sized rocks signal that the volcano was nearing another major eruption. You need to concern these issues before you decide to carry on volcano investigation: Where are there active volcanoes?

Mark their location on a world map, list recent eruptions for each.

Did you find any undersea volcanoes?

Are there any volcanoes that rise from the bottom of the ocean?

How to research the techniques volcanologists to use to monitor volcanoes and to predict eruptions?

Where do the heat, gas and magma of volcanoes come from?

Why do some volcanoes explode when others simple spew out streams of lave?

What are the different types of volcanoes?

Hence, you need to evaluate whether this volcanoes research is value to your scientific research for future similar volcanoes research development. Do you have enough preparation to solve any danger when you stay on the volcano places? Do you have any techniques to use to monitor volcanoes and to predict eruption ? Is your volcano investigation more important to compare other prior scientists' similar volcano investigation. Hence, you can know whether what your volcano investigation aim is clearly and compare which kinds of technique(s) which is/are more suitable to use to monitor volcanoes and to predict eruption more easy.

Hence, you need to own natural environment science knowledge and concept before you want to do any earth scientific research. Environment science can be explained such as, it is the natural environment encompasses all living things and non-living things occurring naturally. The term is most often applied to the Earth or some part of Earth. This environment encompasses the interaction of all living species, climate, weather, and natural resources that affect human survival and economic activity.

The concept of the natural environment can be distinguished by components: Complete ecological units that function as natural systems without massive civilized human intervention, including all vegetation, microorganisms, soil, rocks, atmosphere, and natural phenomena that occur within their boundaries and their nature. Universal natural resources and physical phenomena that lack clear-cut boundaries, such as air, water, and climate, as well as energy, radiation, electric charge, and magnetism, not originating from civilized human activity. In contrast to the natural environment is the built environment. In such areas where man has fundamentally transformed landscapes such as urban settings and agricultural land conversion, the natural environment is greatly modified into a simplified human environment. Even events which seem less extreme such as hydroelectric dam construction, or photovoltaic system construction in the desert, modify the natural environment into an artificial one. However, it is difficult to find absolutely natural environments on Earth, and naturalness usually varies in a continuum, from 100% natural in one extreme to 0% natural in the other. More precisely, we can consider the different aspects or components of an environment, and see that their degree of naturalness is not uniform. If, for instance, in an agricultural field, the composition and the structure of its soil are similar to those of an undisturbed forest soil, but the structure is quite different. However, you also need have human behavioral scientific knowledge to do earth scientific research because earth life history and human life history are same the long time. Hence, if you can apply human behavioral science to earth science to attempt to find this volcanoes life history more easy.

Behavioral modernity is a suite of behavioral and cognitive traits that distinguishes current Homo sapiens from other anatomically modern humans, hominins, and primates. Although often debated, most scholars agree that modern human behavior can be characterized by abstract thinking, planning depth, symbolic behavior (e.g. art, ornamentation, music), exploitation of large game, and blade technology, among others. Underlying these behaviors and technological innovations are cognitive and cultural foundations that have been documented experimentally and ethnographically. Some of these human universal patterns are cumulative cultural adaptation, social norms, language, cooperative breeding, and extensive help and cooperation beyond close kin.

Arising from differences in the archaeological record, a debate continues as to whether anatomically modern humans were behaviorally modern as well. There are many theories on the evolution of behavioral modernity. These generally fall into two camps: gradualist and cognitive approaches. The Later Upper Paleolithic Model refers to the idea that modern human behavior arose through cognitive, genetic changes abruptly around 40,000–50,000 years ago. Other models focus on how modern human behavior may have arisen through gradual steps; the archaeological signatures of such behavior only appearing through demographic or subsistence-based changes.

Archaeological Evidence

In order to classify what traits should be included in modern human behavior, it is necessary to define behaviors that are universal among living human groups. Some examples of these human universals are abstract thought, planning, trade, cooperative labor, body decoration, control and use of fire. Along with these traits, humans possess a heavy reliance on social learning. This cumulative cultural change or cultural "ratchet" separates human culture from social learning in animals. As well, a reliance on social learning may be responsible in part for humans' rapid adaptation to many environments outside of Africa. Since cultural universals are found in all cultures including some of the most isolated indigenous groups, these traits must have evolved or have been invented in Africa prior to the exodus. Archaeologically a number of empirical traits have been used as indicators of modern human behavior. While these are often debated a few are generally agreed upon. Archaeological evidence of behavioral modernity are: burial, fishing, figurative art (cave paintings, petroglyphs, figurines), systematic use of pigment (such as ochre) and jewelry for decoration or self-ornamentation. Besides, Archaeological evidence of using bone material for tools, transport of resources over long distances.

Several critiques have been placed against the traditional concept of behavioral modernity, both methodologically and philosophically. Some researchers argue that a greater emphasis should be placed on identifying only those artifacts which are unquestionably, or purely, symbolic as a metric for modern human behavior.

Hence, if you can know what the human life history in the volcanoes place. You can judge whether the volcanoes place life history in earth or the human life history which is longer time, then you can judge the volcanoes age more easy.

Ocean animal biotechnology scientist case:

Supposing you are ocean animal biotechnology scientist in marine biology. You need to deal with whale hearing and anatomy and with sonar technology. You hope to answer these questions after you decide to do this scientific research.

Are whales the only animals to use sonar for locating objects?
What is sonar? What do they do with?
How does a whale heat them once an animal is dead?
How can a scientist hope to figure out?
How and what the animal hear?

Hence, you need have ethic to prepare to do this research, such as that you need how much time to get answers to solve these challenges, you have how much chance to research successfully, how you would change your research methods if you researched fail, your organization will spend how much salaries and labours and investment to assist your research. Because you need to give feedback to let your organization to know your result to let which feel it is valuable to invest to your research. Hence, your morality is the most important to decide to implement this research to find how and what whale hear etc. ability questions. You need to spend time to attempt to find different methods to attempt to carry on implementing this research to get the result more ease and efficient and effective. This is your moral responsibility.

Supposing you are one ocean animal biotechnology scientist to carry on great white sharks researching.
You need to do work in choppy ocean waters infested with great white sharks. Your job need to lure the sharks closer. You need to concern these issues before you carry on great white sharks research.

How would you feel if you were trying to study twenty five foot long monster shark as it circles closer and closer studying you for its next meal? Should you work to save great white sharks or let them become extinct? Search for arguments on both sides of this question. Why would scientists want to study and understand great white sharks ? List as many reasons as you can. Diving with sharks is dangerous . How would you decide how much danger is acceptable during you own research ? How do you decide how much risk and danger you will accept in your daily life?

Hence, you need have ethic to prepare to do this research. You need to evaluate whether what the influences were to human if these great white sharks became extinct, this research can give what benefits to human, whether your life is more important or great white sharks extinction is more important and you would be immoral if you could not reduce big white sharks death. Hence, it is personal ethic issue to concern whether your research is valuable to attempt or not. You need to spend more time to give recommendations how to save great white shark or let them become extinct. Hence, your recommendations are very important before you decide to carry on researching. Because if you recommendations are useful, then you can reduce more time and danger to work in choppy ocean waters and you can have more confidence to decide whether there are any methods or none methods to save big white shark in possible.

Supposing you need to carry on researching in the South pacific ocean. You aim to research poisonous things in the ocean which fishes and marine mammals are poisonous. You need to concern these issues , such as
Which fishes and marine mammals are poisonous?
Which plants are?
Which ones are poisonous part of the what makes them poisonous?

Hence, you need have ethic to prepare to do this research. Whether you will have chance to cause ill ,even death if the poisonous things can influence to human bodies to cause bad health when who swim in South pacific ocean. If this poison can cause your bodies to be bad health unlucky, you have any medicine to kill this poison immediately. Do you feel what cause poisonous things in the ocean? Can you feel that you can find the poisonous things in the ocean successfully? How much time and how many times do you spend to find the poisonous things? How would you solve the poisonous things if you found the poisonous things? Hence, you need to judge whether it is immoral if the poisonous things could not reduced their numbers to kill some fishes in this ocean. If you can attempt to give recommendations to find any factors to cause poisonous things in the ocean which fishes and marine mammals are poisonous before you do this research. Then, you shall have more confidence to find whether there are none or any factors(s) to cause poisonous things in the ocean. Hence, it is our moral responsibility.

Supposing you need to do deep ocean adventurous research to find the different kinds of new discovery of unique fishes in this deep water area by one flight aviator. You need to know these issues.

How fast does the deep flight aviator travel?

Hence, you need to have this basic knowledge before you begin to enter

deep ocean.
The deep flight aviator's underwater speed can best be compared to an aircraft flying at 330 knots-the average speed for a small jet. The deep flight aviator will do approximately 10 knots. By comparison, whales and other large animals move at around 6 knots.

How deep will go?
Maximum depth for the flight school will be 1,000 feet. initial training drives will be at drive depths (under 150 feet) where we will have driver support, good natural light and will see animal life and shipwrecks (suck by dive operators). More advanced dives will go deeper, according to the customer's comfort and wishes.

What will you see?
0-100 feet-looks like what you see on television (blue water, lots of animals, reefs).
100-300 feet-getting dark (200) and dark at 300. Water will be black and will be illuminated by sub's powerful lights.
500 feet-no natural light completely dark/black. The appearance of the bottom will change from reef to deep ocean. Very few people have been to 500 feet and below. At these depths, 50% of the eyeballs looking back at you are uncatalogued by science.
1,000 feet-alien environment, anywhere you go , you will be seeing a piece of the earth no human has seen before. There will be good visibility.

Hence, you need have ethic to prepare to do this research. How much investment do your organization spend to your research? Do you have confidence to find any unique new kinds of fishes in this ocean? If you can not find any unique new kinds of fishes, how you give feedback to your organization's any enquiry. Did you have adventurous mind or attitude to solve any challenges when your flight aviator had fallen down the 1,000 feet-alien environment in the deep ocean ? Hence, you need to evaluate whether your adventure is value to attempt to do this research. It is your moral responsibility to gather data about the prior discovery of different kinds of unique fishes in deep flight aviator. Hence, you can know whether what the chance is to find any kind of new discovery unique fishes in deep water. Then, you need to let your organization to know whether it is possible or is not possible to find any kind of new discovery unique fishes in this deep water area honestly. If you estimated that it is no chance to find any kind of new discovery unique fishes in this deep water area. Then, you ought to judge where the another deep water area is more easy to find any new kind of new discovery unique fishes and let your organization to know. Till to you can find the deep water area where you feel it is more easy to find any new kind of new discovery unique fishes because it is your moral responsibility to your organization and your research job.

CHAPTER SEVEN

How scientist ethic influences business environment development

Nowadays, if our traditional business social environment has no ethic or moral regualation to prohibit any businessmen or business industry leaders or professionals or founders to do any immoral behaviors legally, e.g. accountants, lawyers, managers etc. Then, will our business social environment become become better or worse? For example, business professionals have taken the products of science and revolutionized the fields of agricultural, transportation and medicine. Busines professionals have taken the products of act and dramatically increased our access to them. If agricultural scientists intend to produce or invent to increase any new agricultural fruit or vegetable food number, but they apply the harmful human health chemical materials to increase agricultural vegetble or fruit number or improve their taste to be more better to be eatten. The, it will cause human get illness or diseases attack, due to human often eat this new scientific vegetable or fruit food. For medicine scientists only consider to increase new medicine number to satisfy patient needs. But, they have not attempted to do any enough experiments to test to confirm these new medicine whether they are dealted patient individual health successfully, even whether they have negative unhealth influence to cause patient individual body to be worse, or even cause another illnesses (diseases) after they eat the new medicine. So , it seems that every business professional's immoral behavior will bring human's life threats, concerning the food or medicine or soft drink untest DNA invention. Hence, every country's government needs have a good understanding of how protecting some business professional immoral business behaviors in order to avoid or reduce the consumption threats of areas of our daily life.

I feel the core of business moral jobs in our societies include: education, science, art, law, doctor, medicine, engineer, accountant, architect etc. different professionals. The reasons are that the profession of science creates value: the discovery of new knowledge, the profession of art creates value, objects that express and evoke imprtant human themes. However, each profession, some individuals act unethically, the reasons are possible because self interest pursuit and promotion of selfish.

Hence, the best business environment, it needs have business ethics regulation to prohibit business professionals to do immoral behaviors. Business is assumed to be at best a moral enterprise and the expectation is often that the business practice is more likely than not to be immoral interest. In fact, business is driven by seld interest. For example, some brands of environmentalsim products, that industrial chemicals are poisoning everything function psychologically. As general ,axiomatic truths are concerned. Hence, environmentalist believe the industrial chemical product manufacturers only consider self interest is outside of morality to give only interest in business. So, why do the philosophers traditionally put morality and self interest in different categories? Also, it implies self interest is argued to be major problem in business in two ways. First, the profit motive can lead oe individual to harm another. Second, the profit motive can lead individuals not to help the less fortunate, self interest leads, e.g. the standard to harm other stakeholders, argument against plant relocations is not that the company is harming the rights of the workers, rather since the workers will be in a more desperate situation, the moral company would be willing to forgive the profit opportunities that a plant relocation would offer them. Hence, self interest will be one major factor to cause business immoral behavior occurrence. Another reason is limited resources to cause business immoral behavior occurrence.

Due to in traditional ethics, producing our way out of scaraity is not seen as an option. Scarcity means there is not enough to something, e.g. scarces resources to produce. That the manufcturers' self interest is in fundamental conflict because of economic scarcity. Also due to manufacturers need to compete to win any limited resource. In a world of scarce resources, business is fundamentally to pursue the profit motive to lead competition, the result will cause immoral business behaviors in possible. Hence, illegal black market will be caused by resources scarcity, e.g. black computers, mobiles etc. product illegal market transactions.

Consequently, business environment development will have direct relation with immoral business behavior. Because if human continue to attempt to do any immoral activities to achieve their profit or self interest aim. Thus, our global business environment will not develop easily. All technological products can be improved to satisfy human's daily needs as well as foods won't be safe to be eaten. Finally, positioned food will be serious to cause that we will harm ourselves lifes.

Scientist's social responsibility

I shall indicate that why scientists needs to concern human genetic engineering science ethical and moral issue, it may seem of sience fiction, but genetic engineering has now entered the actual of human possibility. so, it is influenced to any patient's life risk. What does our science actually allow scientists to do today and what could it realistically achieve in the future? What are the true potential benefits and risks of this powerful technology? Such as, DNA transfer and use of cell nuclear transfer scientific moral research. Nuclear DNA health science aims to treat disease to assist healthy children. However, human genetic engineering scientists need raise some important concerns that will need to be addressed before any DNA transfer and use of DNA nuclear transfer scientific disease killed treatment to be entered medicine to any patients' bodies decision that can be made on whether doctors should be pursued more thoroughly.

Although, DNA cell health treatments has potential benefits, but it has also risks of those genetic engineering now. The new genetic engineering technology is DNA transfer technique designed of the purpose of eliminating disease, which can kill unhealthy cells in patients‘ bodies by the healthy DNA cells. For example, disease of heart, liver and kidneys and loss of coordination and muscle weakness. These patients' bad disease cells can be attempted to be killed by the health DNA cells medical drug treatment. However, DNA transfer technology is a attemption of stage, Hence, it has still risk to unsure to enter the health DNA cells to any human's (patient's) bodies. Besides, it has been no offical clinical trial in humans, so trial is illegal.

Hence, DNA trafer cell technology is a test stage, so it has risk to patients' bodies if the DNA is unhealth cells. Then the DNA transfer is caused danfer to the patients more seriously, even death risk is possible to occur to patients. For example, when medicine (grug) scientists are carrying on human researching to serve to ensure the safety of new medicine, establish tolerable exposure levels for environment and workplace hazards and determine the effectiveness of new interventions in public health, education and other fields. Without volunteers, these studies would be impossible to conduct. Recognizing society's responsibility to protect human subjects of reasearch from avoidable harm and unethical treatrment. These rules reflects widely accepted principles of ethics. These principles are rooted in values that find expression in many sources of moral philosophy, theological traditions and codes, regulations and rules. They are ethical science or moral science. Due to medical research that poses risk of physical injury rightly. However, the ethical, moral and social issues involve genetic modification , i.e. the direct manipulation of an animal's genetic make-up genetic modification of animals was first achieved with nice in 1980 year, and of cattle, sheep and pigs by about 1985 year. The reasons indicate genetically modified animals are produced include to help scientists to indentify isolate and characterise genes in order to understand more about their function and regulation to provide research models of human diseases to help develop new drugs and new strategies for repairing defective genes to provide organs. And tissues for use in human transplant surgery to produce milk which contains proteins or to alter the composition of the milk to improve its nutritional value for human infants.

In conclusion, in fact, the responsibilities of scientists, who ought to give the corrective scientific reasoning to apply whose ability to scientific community and the broader society, due to who own the nature of responsibility or role generally. Their level of responsibility has minimum demand or ideal to social benefit and who bears the responsibility to the individual or the community of benefits. I think scientists' occupation have the traditional range of research ehtics, or the responsible conduct of research, it is usually concerned over falsification, fabrication, plagiarism and treatment of human and animal and natural and foods and earth etc. subjects. However, I think scientific moralty requires at least four dimensions of consideration to be rules to scientists' behavior. Such as, a set of bases is for the responsibilites; a distination is in kinds of responsibilities; a distinction is between minimally acceptable and ideal behavior and a distinction is between collective and individual responsibility.

Thus, scientists need to understand their value of science in society and whose duties are standard complexities in accounts of responsibilities, who need to bear responsibility to carry on researching any new challenges. Moreover, scientists ought need to attempt to answer these questions before who decide to attempt to carry on any scientific researching every time. Is it a general responsibility or a role responsibility? Is it a minimum requirement or an idal? The reason is that scientists need are responsible to rather than responsible for. Because responsibility is for raising problems of interpretation because and ambiguities that reside in that expression. We we say that someone is responsible for something, we could mean that who are be held worthy for something or that it is their ongoing job to take care of something. It combined one thing could be responsible for in either of these senses, makes this a poor starting point for analysis. Hence, if seems have moral and legal overlap to any scientists' behavior, due to scientists are responsible for in particular cases or how public might ensure who are held accountable to people or institutions. In our society, legal and moral overlap is such a map, due to society can't examine the responsibilities of scientists to see if the institutions of science are adequately constructed.

What is scientist's social responsibility? Why do scientists need have social responsibilities? What does it influence to society of scientists lack social responsibilities to do any scientific research? Scientists have a long tradition of discussing their responsibilities as a balance between their professional autonomy and their general moral responsibility as human beings (Douglas, 2003).

I believe scientists need have social responsibilities, because if they can do reasonable social responsibility of behaviors, then it represents that they have good moral performance to concern whether their any scientific researchs are only considered themselves benefits for intention, even if their scientifical researchs can harm human's safety. who will neglect on the possible poor effects, due to their immoral scientific invention or research. For example, nuclear invention can be good thing, if it is applied to space flight energy, or earth energy to avoid vehicle oil , or any transportation tool, e.g. gas, oil shortage in the future one day. But if nuclear invention is applied to produce nuclear bomb. Then, it will influence human's life safety. For DNA example, if DNA is applied to detect any criminal behavior cases to find whom are the criminal people had done any crimes . If DNA is applied to medical health aspect, it can bring hosptial patients or police force advantages. If DNA is applied to grow good taste or big size of fruit or vegetable or beef or port etc. different kind of food to provide to human to eat in order to avoid our food shortage challenge occurrence in future one day. It can solve food shortage challenge, but if DNA is applied to reproduce human ourselves or any animals themselves. It is immoral scientific research and it is immoral social responsible behavior to the scientists, although it is not illegal research in nowadays law system.

For animal scientist morality , there are reasons why animal scientists need to concern genetically modified animals are produced

1. To help scientists to identify, isolate and characterise genes in order to understand more about their function and regulation. Genetic modification can be used to knock out the activity of a particular gene. By correlating loss of function with this "knock out" it is possible to gain information about the role of the gene and the product for which it codes.

To provide research models of human diseases, to help develop new drugs and new strategies for repairing defective genes ("gene therapy").

Animal models of diseases have been used for many years by exploiting naturally occurring mutations in genes, and in-breeding laboratory strains of animals carrying the mutation. An example is a mouse model of Duchenne Muscular Dystrophy. Genetic modification has been used to produce animal models of many diseases including mice with predisposition to cancers, and mice with cystic fibrosis. These models may be made by "knocking out" the activity of genes, as described above, or by inserting defective genes.

By inserting additional copies of a gene into laboratory mice and observing the effects, scientists have recently confirmed the role of this gene in a disorder of human babies that is associated with increased susceptibility to childhood cancers. This will aid the design of new medical treatments.

For nuclear scientist example, if the nuclear scientist neglects to consider morality to research nuclear radiological weapons of mass destruction to manufacture. It will encourage to cause war occurrence difference of different counties in the future. For example, a weapon of mass destruction (WMD) is a nuclear, radiological, chemical, biological or other weapon that can kill and bring significant harm to a large number of humans or cause great damage to human-made structures (e.g. buildings), natural structures (e.g. mountains), or the biosphere. The scope and application of the term

has evolved and been disputed, often signifying more politically than technically. Originally coined in reference to aerial bombing with chemical explosives, since World War II it has come to refer to large-scale weaponry of other technologies, such as chemical, biological, radiological, or nuclear. Who can think at this present time without a sickening of the heart of the appalling slaughter, the suffering, the manifold misery brought by war to Spain and to China? Who can think without horror of what another widespread war would mean, waged as it would be with all the new weapons of mass destruction?
At the time, the United States (with help from Western Allies) had yet to develop and use nuclear weapons. Japan conducted research on biological weapons and chemical weapons had seen wide battlefield use in World War I. "It is a very far reaching control which would eliminate the rivalry between nations in this field, which would prevent the surreptitious arming of one nation against another, which would provide some cushion of time before atomic attack, and presumably therefore before any attack with weapons of mass destruction, and which would go a long way toward removing atomic energy at least as a source of conflict between the powers. Nowadays, chemical weapon war will also caused if any biological scientists neglect who morality to use their biological knowledge to manufacture any new kinds of chemical weapons to sell to different countries to gain income to achieve rich goals. For example, after the 11 September 2001 attacks and the 2001 anthrax attacks in the United States, an increased fear of nonconventional weapons and asymmetrical warfare took hold in many countries. The most widely used definition of "weapons of mass destruction" is that of nuclear, biological, or chemical weapons (NBC) .

However, there is an argument that nuclear and biological weapons do not belong in the same category as chemical and "dirty bomb" radiological weapons, which have limited destructive potential (and close to none, as far as property is concerned), whereas nuclear and biological weapons have the unique ability to kill large numbers of people with very small amounts of material, and thus could be said to belong in a class by themselves. Such as, Nuclear weapons or nuclear-weapons-usable material or any sub-systems or components or any research, development, support or manufacturing facilities relating to [nuclear weapons]. Chemical and biological weapons and all stocks of agents and all related subsystems and components and all research, development, support and manufacturing facilities. Hence , you are one energy scientist, you have moral responsibility to judge whether your new energy invention would bring more benefits to human or would bring more damage or harm to human if your new energy invention was achieved successfully.

Consequently it bring this question: why scientists need have social responsibilities, it is due to that scientists and our societies have a particular relationship between them as well as science and our society is necessary in order for science to be responsible and also that scientists need to conduct their science within the structure of this relationship proper. For this case example, the Manhattan projecy, the development of the atomic bomb in the US, during World War II, which led a number of occupation (Rhodes , 2012). Thus, it implies the atomic bomb invention scientists will feel who had social responsibilities, how to influence human's life safety in different countries. If some countries applied atomic bombs to be weapons to attack the countries to achieve control or management ambition to these lacking high technological weapons of countries.

Consequently, in micro social responsibility view point, it seems that scientists immoral behaviors will only influence to themselves country's people independently. Otherview, in macro social responsibility view point, it seems scientists immoral behaviors will influence themselves country and other countries people, even global people whose life safety. Such as the World War II occurrence case.

However, some scientists argue why who need have social responsibility, because they invent any scientific products. It means that they give attribution or welfare or benefits to human absolutely. I can give this example to support their view point. Although, their contribution rationality articulates science as a societal institution akin to the healrh care and the educative system. It is part of society and serves certain societal goals: Clearly, the aims of science , particularly in the case of the biomedical science are closely linked to certain ethical, social, or political goals (De Melo-Martin, 2008,39).

Hence, although for doctors are doing medical attribution to patients, but if they lack social responsibility of mind to do immoral behavior, such as earning expensive medical service income to help female clients to do any health skin or beautiful facial medical service. But they had not test this skin or facial medical medicine whethe which won't hurt human's face or skin to cause disease or illnesses during their medical processing. It is danger to the facial clients' life safety. It brings this question: Whether the facial doctor's behaviors are either beneficial to their facial clients to let them to get beautiful faces or health skins or hurt to their facial clients skins or faces. Because it seems that the facial doctors are moral and beneficial to their facial clients. But in fact, their behaviors will possible hurt their facial clients' faces or skins , even causes their life safety. Hence, as any indication that why any scientists need have social responsibilities before they decided to make any behaviors.

This rationality often articulates faith in education as a means to make scientists more morally: " What science education now requires is meta science, a discipline that extends beyond conventional, philosophy and ethics to include the social and humanistic aspects of the scientific enterprise. For example, students need to learn about th societal responsibilities of research scientists, and to rehearse in advance same of the moral responsibilities that they are likely to meet "(Ziman 2001, 165). Hence, to teach how scientists believe moral and social responsibility issue is important to themselves. They must need to be educted to learn what moral or social responsibility who need have from universities. So, I suggest that universities ought have this science moralty subject to be taught to any science subject students to prepare them to do any moral behaviors in their future career.

I shall explain why scientists need have moral conduct as below reasons:
First, any professionals must have professional moralty to prohibit who do any conducts which damage social benefits, e.g. accountants need have morality to avoid to do any false accounting recrd to ern social benefits, lawyers need have morality to avoid to do any unfair criminal conclusion to earn unfair or unreasonable legal service charges, doctors need have morality to avoid to give high life risk and cheap price and poor quality of medial medicine to treat the poor patients to eat to cause they have life dangers, due to they have no enough money to pay expensive medical fees and get the unfair poor medical treatment, businessmen need have business morality to do any legal buying and selling transaction behaviors. Also, different scientificial professionals need have scientific professional morality. For example, engineers and architects need have moral responsibilities to design any building properties or machines drawing plans

in order to acheive the buildings are safe to let people to live or work, or the machines are safe to let people to use. Hence, in our societies, any professional occupations need have professional moral regulation to prohibit any professional occupation employers to do immoral professional behaviors. So, it brings this question: Why have scientist occupations no any moral regulation to prohibit them to do any immoral behaviors in our societies? Is it unfair to other professional occupations is scientists have no any professional moral regulation to prohibit them to do immoral behaviors to carry on their scientfic researching? Hence, it is only possible if science allows social concerns to form part of the scientific process, instead of focusing only on technical aspects. In this way, the integration rationality articulating from the contribution rationality by articulating that any scientific knowledge production is a moral production and responsibility is something that develops through any scientific research experimental process.

In conclusion, any scientists need have social or moral responsibilities to a large group of different stakeholders from outside the scientific system should be involved in the conduct of science. In these rationalities, it seems that social scientists can enter the laboratory and co-develop knowledge to do any scientific experiments morally. On the one hand, the demoration rationality focuses on optimizing the scientific process befire the fact by installing a strict moral code among the scientists that should be focuses on honestry and accuracy in their work. On the other hand, their scientific contributions will bea necessary source of moral knowledge about how to develop in a good society from which science needs to learn.

What is scientific morality and scientists relationship?

i. What is scientific research meaning?

Scientific research is a systematic investigation to establish facts. An attempt to find out something in a systematic and scientific manner. A systematic investigation designed to develop knowledge and a focused systematic study undertaken to increase new knowledge or understanding. It is the collection of information about a particular subject.

ii. Why scientists need to concern morality ?

An enquiry that involves seeking evidence to increase knowledge. For a biomedical scientist example, who needs to share between different research disciplines, such as the need for some methodology, will be interpreted in significantly different ways. His research aimed to demonstrate the phenomenon of human conditioning by conditioning an 11 months old infant to fear rats by associating, then with fear inducing circumstances, such as a loud noise; biomedical experiments include freezing to induce hypothermia, infection of research subjects with malaria and tuberculosis (TB) and many consent of the research subjects and often leading predictably or extreme gain, mutilation and death. This issues are unmoral bad result. Hence, biomedical scientists need to concern whether whose scientific research is moral to society.

iii. How to teach scientific morality to students?

For another example, biological scientific research, a mixed method design was to address the issue of effectiveness of ethical frameworks in enabling students to develop ethical reasoning skills in year 10 biotechnology program. This ten weeks program, focused a gene technology, genetically modified foods, genetic engineering and reproductive technologies. Each student attended to do experimental design quantitative data from the pre and post program questionnaires were used to determine the effectiveness in the use of the ethical frameworks. The questionnaires assessed the student's understanding and ethical thinking, attitude and opinions of biology scientific knowledge and ended with a section on the student's religious faith.

iv. What is morality meaning?

Morality means manner, character, proper behavior" to be judged whether the scientist's behavior is the differentiation of intentions, decisions, and actions between those that are distinguished as proper and those that are improper by the acceptable standard of the society. Morality can be a body of standards or principles derived from a code of conduct from a particular philosophy, religion, or culture, or it can derive from a standard that a person believes should be universal. Morality may also be specifically synonymous with "goodness" or "rightness." Moral philosophy includes moral ontology, or the origin of morals, as well as moral epistemology, or knowledge about morals. Different systems of expressing morality have been proposed, including deontological ethical systems which adhere to a set of established rules, and normative ethical systems which consider the merits of actions themselves. However, immorality is the active opposition to morality (i.e. opposition to that which is good or right), while amorality is variously defined as an unawareness of, indifference toward, or disbelief in any set of moral standards or principles.

v. What is scientific morality and medical professionals relationship?

Beauchamp T. L., & Childress J.F. (2001) defined that "ethic is the moral reasoning of actions. For example , Why medical ethics is important? Medical professionals increasingly find themselves confronted with moral questions, e.g. ethic guideline address special medical services , such as blood transfusions and health services and health care for patients living with HIV/AIDS ethical issues that arise a clinical medicine. It addresses justice, equity and access to medical care. It also focuses on general duties of doctors, dentists and pharmacists."

For medical scientific research and ethic relationship, the global medical profession has maintained simple ethical standard for more than 4,000 years. the majority patients, physicians and other medical and health care providers often face ethical challenges, many of them consider ethics to be a concept pertaining to the avoidance and biology is referred to as bio ethics or biomedical ethics. Ethics is an intrinsic part of medical practice and shapes the medical profession and it must be implemented and individual behaviour of approaches that are based on someone's beliefs. Medical ethics isn't about avoiding harm, rather it is a set of norms, values and principles. These norms, values and principles are intended to govern medical ethic conduct.

vi. Why scientists need to know ethics.

To answer this question, you need to attempt to answer these questions before you find the reason to answer this main question, such as below:

What is it to live a morally scientific research life?

Why is scientific morality important to carrying on researching?

Are moral principles valid only as scientists depend on their countries' cultural approval or are there universal moral truths?

How should scientists live in moral life?

Are there intrinsic values?

Which is the best moral scientific theory when who are carrying on researching?

Can scientists derive moral values from facts?

Is there a right answer to every scientific problem in life?

Which is the relationship of religion to morality?

Thus, I believe ethics in science is basic principles. In fact, science has a special role with respect to ethic, society demands high standards of scientists, it isn't always easy to determine the right thing to do, breaches of scientific ethics make headlines and ruin careers. Ethics have different categories. They include as below:

● Personal ethics: Morality .
Professional ethics: Standards & expectation.
Societal ethics: law.
● What is ethical standards? Should do the right thing? Should I do the right thing? What is the right framework for making ethical decisions?

As Gert., (1988) explained " ethic is a system of public, general rules for guiding human conduct. Ethics (also known as moral philosophy) is the branch of philosophy which addresses questions of morality. The word "ethics" is commonly used interchangeably with morality, and sometimes it is used more narrowly to mean the moral principles of a particular tradition, group, or individual. Likewise,certain types of ethical theories, especially deontological ethics, sometimes distinguish between ethics and morals."

vii. Why do scientists need to know whether their behaviour are be acceptable to morality ?

I believe that science means the research is for truth, such as a quest for objective knowledge about nature; it is social institution means serving society's needs and improving people's lives as well as it made up of people with human needs and desire; and it is a profession, involving training, standards , a career, respect and privilege. Whether where is science done? Science is done at universities, in government, laboratories or government funding, in military laboratories or with military funding, industry laboratories or with industry funding. Scientists need to know different standards and goals apply to organizations under different circumstances. University goal is to advance knowledge to educate students to serve the public, government goal is to protect the nation's people and property to compete and cooperate with other nations to keep global security and to serve the companies or general industry, and increase opportunities in the global market. Hence, when any scientist works in different organization, then the organization will has different goal and the scientist will have the different goal and different scientific morality or ethic in different situation. Science include Astronomy, Biology, Botany, Computer science, Chemistry, Cosmology, Geography, Geology, Jurisprudence, Mathematics, Paleontology, Physics, Economics and others etc. different scientific researches.

A scientist is a person engaging in a systematic activity to acquire knowledge that describes and predicts the natural world. In a more restricted sense, a scientist may refer to an individual who uses the scientific method. The person may be an expert in one or more areas of science. Also scientists perform research toward a more comprehensive understanding of nature, including physical, mathematical and social realms. Hence, Scientists are also distinct from engineers, those who design, build, and maintain devices for particular situations; however, no engineer attains that title without significant study of science and the scientific method. When science is done with a goal toward practical utility, it is called applied science. An applied scientist may not be designing something in particular, but rather is conducting research with the aim of developing new technologies and practical methods. When science is done with an inclusion of intangible aspects of reality it is called natural philosophy. Scientists are also distinct from engineers, those who design, build, and maintain devices for particular situations; however, no engineer attains that title without significant study of science and the scientific method. When science is done with a goal toward practical utility, it is called applied science. An applied scientist may not be designing something in particular, but rather is conducting research with the aim of developing new technologies and practical methods. When science is done with an inclusion of intangible aspects of reality it is called natural philosophy. Besides, Science and technology have continually modified human existence through the engineering process. As a profession the scientist of today is widely recognized. Scientists include theoreticians who mainly develop new models to explain existing data and predict new results, and experimentalists who mainly test models by making measurements — though in practice the division between these activities is not clear-cut, and many scientists perform both tasks. However, scientists can be motivated in several ways. Many have a desire to understand why the world is as we see it and how it came to be. They exhibit a strong curiosity about reality. Other motivations are recognition by their peers and prestige, or the desire to apply scientific knowledge for the benefit of people's health, the nations, the world, nature or industries (academic scientist and industrial scientist).

Scientists also tend to be less motivated by direct financial reward for their work than other careers. As a result, scientific researchers often accept lower average salaries when compared with many other professions which require a similar amount of training and qualification. Hence, scientists are professionals, whose mission need to invent any new methods to solve any challenge, as human are encountering. It implies that who must need morality to do their scientific research to achieve benefits to human, but who can't earn extra awards unfairly from social assistance. Hence, scientists must need to know whether their behaviour are be acceptable to morality before or during they decide to do any scientific research.

CHAPTER EIGHT

Rethinking morality in management science

What is morality in management science? Why does management need to concern moral behavior? Because most management researchers feel whose behaviors are possible right, but in moral view point, whose behaviors are possible wrong. This is a subjective personal moral judgement to management researcher individual behavior in any organization. However, if the management researchers can know how to judge whose behaviors whether ar more moral reasons to be accepted in societies . The, who can judge how to choose to do more right moral behavior more correctly in any organizations.

Some management scientists should not assume that one value can be objectively better than another, or that any values are objectively right or objectively wrong. So it brings this question: Are values objective? Are there sometime objective by god reasons to moral actions? Perhaps it is time to rethink hesitation about accepting moral objectivity in management search. Usually management researchers choose to decide to do immoral management behaviors. It is possible that who is seemed to have effort to help whose organization to gain benefits and dominant economy. So, in economic view point, the management researchers are helping their organizations to gain, it is possible that the management researchers do not feel who are doing immoral behaviors to assist their organizational development. Also, even whose behaviors are legal, but it is not represent their behaviors must be morality in societies.
Hence, knowing how to judge whether it is moral behaviors, it is important to every management research in nowadays organizational environment. In any organizations, that agency theory fails to predict corporate

performance, because of mechanisms of monitoring, independence and incentives (Dalton et al. 2003: Dalton et al. 1999) . Due to this reason of self-interested, so management level staffs (researchers) often do immoral behaviors to influence whose whole organizations to bring negative effect to publicity. Because management researchers will feel themselves self-interest is important than whole organizational interest. It is self interested immoral behaviors, although it is not represent themselves behaviors must be illegal in society.

However, good explanations of moral / immoral behvior and how management researchers are affected by moral principles are ones that sometimes make use of the objective structure of moral reasoning. This relates to equate what ought to be with what " is". But is is to say that because management researchers think in ways about ethics, those patterns may be revealed in their actual behavior. Take a simple example related the value of fairness, e.g. the management researcher feels unfair compensation or welfare treatment or salary is provided from whose employer. So, it influence who choose to do immoral activities in whose organizations. So, it seems unfairness is a factor to cause management researcher's immoral behavior or the another factor of perhaps management researchers' hesitation to refer to moral principles when explaining behavior is from the assumption that moral reasons or principles can never serve as causal explanations of behavior.

Anyway , in any organizations, permission and obligation in ethics necessity and possibility in logic principle is essential to be prohibited to every management researcher behavior. The reason is because just or unjust (fair or unfair) behaviors are caused by economical self-interest of human behavior (management researcher behavior). Also, the prior organizational fair researchers are mostly explored relationships between fair or unfair behaviors in the one hand, and self-seeking motives on the other hand.
Consequently, I conclude that general management researchers choose to decide to do any kind of immoral behaviours in organizations. The reasons are usually due to that who feel unfair welfare or compansation treatment or economic self-interest motive factor to influence tham to choose to do unethic or immoral behaviors in their organizations. Thus, employers need to consider how organizational policies are applied prohibit management researchers choose to do any unethic or immoral behaviors in any suitations

in organizational environment easily.

3.1 How common morality relates to business and professions?

Artificial intelligent scientist moralty

Supposing you are one mathematician/computer scientist in first mass computer laboratory. You need to design the root language and operate systems for the different computer languages. The first mass produced commercially available computer in world . Besides writing programs, you was a true pioneer in designing and developing whole languages for computers including the most successful languages for computers general purpose. You need to concern these issues:

How is a computer different then a calculator?

What is a computer language?

How does a computer language work?

How does it let us talk to machine?

How can you create a computer language?

Jurisprudence and mathematics are often grouped with the sciences. Some of the greatest physicists have also been creative mathematicians and lawyers. There is a continuum from the most theoretical to the most empirical scientists with no distinct boundaries. In terms of personality, interests, training and professional activity, there is little difference between applied mathematicians and theoretical physicists.

Computer ethics is a part of practical philosophy which concerns with how computing professionals should make decisions regarding professional and social conduct.

Computer morality can include such as:

a. The individual's own personal code.

b. Any informal code of ethical conduct
that exists in the work place.

c. Exposure to formal codes of ethics.

To understand the foundation of computer ethics, it is important to look into the different schools of biology ethical theory. Each school of ethics influences a situation in a certain direction and pushes the final outcome of ethical theory Relativism is the belief that there are no universal moral norms of right and wrong. In the school of relativistic ethical belief, ethicists divide it into three connected but different structures, subject (Moral) and culture (Anthropological). Moral relativism is the idea that each person decides what is right and wrong for them. Anthropological relativism is the concept of right and wrong is decided by a society's actual moral belief structure. Deontology is the belief that people's actions are to be guided by moral laws, and that these moral laws are universal. Utilitarianism is the belief that if an action is good it benefits someone and an action is bad if it harms someone.

This ethical belief can be broken down into two different schools, Act Utilitarianism and Rule Utilitarianism. Act Utilitarianism is the belief that an action is good if its overall effect is to produce more happiness than unhappiness. Rule Utilitarianism is the belief that we should adopt a moral rule and if followed by everybody, would lead to a greater level of overall happiness. Social contract is the concept that for a society to arise and maintain order, a morality based set of rules must be agreed upon. Social contract theory has influenced modern government and is heavily involved with societal law. Hence, you need have computer scientific morality to decide whether who are/is your new computer language invention's customer(s) group(s) and how who will apply your new computer language to influence society to bring harms or benefits more in the future.

I think moral issues concern the environment, specially the treatment of animals, and such issues as abortion and euthanasia, because these scientists often need to make moral decisions and judgement to make any experiments. The uncontroversial nature of those matters is shown by every scientist's lack of hesitancy in making negative moral judgements about whose who harm others simply because they do not like them. It is shown by the same lack of hesitancy in making moral judgements, unfair deception, breaking of promises, cheating, disobeying the law and not doing one's duty. For example, doctors choose animals' bodies to carry on killing any cancer cells experiments. It is immoral behavior to any animals, which have not dead. Because they used to be attacked by any kind of cancer cells and than doctors will give any new medicines to treat their bodies to attempt to test whether the new medicines are effective to treat any cancer cells. Do you feel animals won't feel hurt or painful when which bodies are tested to attack cancer cells by any new medicine experiments? If it had no any new medicines to be confirmed to attack their cancer cells in these animals' bodies successfully. Then, these animals will fell painful to die, due to any cancer cells experiments. For another example, whether these was a moral difference between a rational refusal of food and fluids by a competent refusal of medical treatment. To this patient's refusal of medical treatment suitation, the doctors need to do moral judgement whether who ought give medical treatment to the competent terminally ill patient or ought not give medical treatment to the competent terminally ill patent. It is too difficult to make moral judgement because the competent terminally ill patient who is probable to feel painful to alive. So, who decides not accept any medical treatment to reduce painful alive. But is the doctor promoted to him/her does not attempt to give any medical treatment to the competent terminally ill patient. It seems that the doctor shortens the patient's life days. Hence, the doctor will feel difficult to make right decision in the medical treatment case, also it is not absolute right or worng answer to the doctor's behavior.

In conclusion, I think any organizations need have rules to control their employees' behaviors, including professionals or general employees. So, the common morality, it includes: rules prohibiting acting or attempting to act in ways that cause, or significantly increase the prohibility of causing, any harms that all rational persons want avoidance. Whereas it is possible to obey the moral rules all of the time impartically with regard to with regard to everyone in our societies.

What is professional ethic? Professional ethics is not distinct from common morality. Rather, a profession takes on certain duties that are not duties for those duties must be compatible with the framework provided by common morality. Nor can any business impose duties on its employees that would be considered morally acceptable by all informed, impartial, rational persons. Also, a job can't impose duties that are morally unacceptable. For example, a driver of a getaway car does not have a duty to help bank robbers escape after a bank robbery, even if who has been paid to do so. An employee of an advertising company does not have a duty to help compose an advertisement that will persuade young people into smoking ot taking any other additive drug. Scientists who are employed by a cigarette company do not have a duty to help make that producing more addictive.

Consequently, it implies professional ethic is more strict to compare to general moral behaviors to general employees behaviors in any organizations. Because professionals need have better ethic to conduct daily behaviors for whose jobs and duties to compare to general employees in organizations. Finally, it also explains that it must have the difference between general morality and professional ethic in our societies nowadays.

However, (AI) technological scientists and other kind of scientists who need have moral or ethic consideration because (AI) consumers and any science product consumers who must concern their products whether their products will bring either safe use or unsafe use from law regulation prohibition.

CHAPTER NINE

Robots ethic how influence human behavioral change

The integration of robotics and artificial intelligence into the global economy necessitates a rigorous examination of moral and ethical frameworks. My analysis of robot ethics, the imperative for business owners and scientists to consider these factors stems from the fundamental shift in the nature of agency and responsibility. When manufacturers and researchers develop autonomous entities, they are not merely creating tools; they are creating systems that interact with human society in ways that can alter legal, social, and moral landscapes.

The necessity for business owners and scientists to prioritize robot ethics is rooted in the "moral weight" of the choices made during the design phase. Scientists must consider the ethical implications of their research because the development of artificial intelligence raises critical questions regarding the moral status of these entities and the responsibilities humans hold toward them. For business owners, the integration of AI is not just a matter of efficiency but a matter of liability and societal impact. If a robot is permitted to enter into contracts or perform tasks that affect human well-being, the business owner must account for the moral consequences of those actions, as the robot may not possess the capacity to be held fully responsible for its decisions.

The factors involved when these entities neglect to consider robot ethics are profound and multifaceted. First, there is the risk of "moral vacuum," where the lack of clear ethical guidelines leads to unpredictable outcomes that can result in significant legal and financial liability. Second, neglecting ethics can lead to a loss of public trust, which is a critical asset for any business or scientific institution. As noted in broader literature on the subject, the failure to address the "black box" nature of AI decision-making can lead to systemic biases and discriminatory practices that violate social norms and legal standards.

Furthermore, from a philosophical and economic perspective, failing to define the moral status of robots can lead to a crisis of accountability, where it becomes unclear whether the human creator or the machine itself bears the burden of a negative outcome.

Ultimately, the ethical development of robotics requires a transition from viewing AI as a purely technical challenge to viewing it as a socio-moral one. Scientists and business owners who ignore these dimensions risk creating systems that are technically proficient but socially destructive, potentially leading to long-term economic instability and the erosion of the ethical standards that govern human interaction.

In fact, The question regarding the specific consequences for businesses that ignore robot ethics as articulated by my opinion presents a significant challenge, authoritative literature regarding corporate social responsibility, algorithmic accountability, and the socio-economic impacts of artificial intelligence.

The consequences for businesses that neglect robot ethics are extensively detailed in foundational texts on the intersection of technology and business strategy. Scholars such as Virginia Eubanks have demonstrated that when organizations deploy automated systems without rigorous ethical oversight, they risk "shattering the social safety net" and "intensifying discrimination," which leads to severe reputational damage and legal liability. Businesses that fail to integrate ethical considerations into their AI deployment often face "automation anxiety" among their workforce, which can manifest as organized labor resistance, loss of institutional knowledge due to skill atrophy, and a decline in employee morale.

Furthermore, the failure to prioritize ethical AI—specifically regarding transparency and human-in-the-loop decision-making—exposes firms to regulatory sanctions, such as those outlined under the European Union's General Data Protection Regulation (GDPR), which penalizes decisions made solely by automated processing without human oversight.[From a strategic management perspective, ignoring these ethical dimensions can result in the "atrophy of core skills" within the organization, where reliance on black-box algorithms replaces the critical human judgment necessary for long-term innovation and crisis management. Ultimately, businesses that treat ethics as an afterthought risk creating systems that are not only biased and exclusionary but also fundamentally incompatible with the human-centric values required for sustainable market participation.

The integration of artificial intelligence and robotics into the corporate landscape necessitates a rigorous examination of moral and ethical frameworks. According to John Lok in his work Robot Ethic, businesses that neglect the ethical dimensions of AI development face significant risks that extend beyond mere technical failure.[1] When corporations prioritize efficiency or profit without establishing moral limitations, they risk creating systems that lack accountability, potentially leading to legal liabilities and a breakdown in public trust.

My view emphasizes that the consequences for businesses ignoring robot ethics are multifaceted. First, there is the issue of moral responsibility regarding the "well-being of human" and the societal impact of automated labor. If a business deploys autonomous systems without considering the ethical employment of these entities or their impact on human workers, they may face severe reputational damage and social backlash.

Furthermore, my view argues that businesses must determine whether their robotic entities are "permitted to enter contracts" and to what extent these entities can be held "fully responsible" for their actions. By failing to address these legal and moral boundaries, companies leave themselves vulnerable to complex litigation where the "moral weight of the choice" becomes a point of contention in courtrooms and public discourse.

Ultimately, my view suggests that ignoring these ethics transforms a business decision into a moral hazard, where the lack of a clear ethical framework makes it impossible to assign liability when a robot's actions result in negative consequences.

In the discourse surrounding the integration of artificial intelligence and robotics into the corporate sphere, my view emphasizes a critical bifurcation in the analysis of these technologies. According to his research, the ethical implications of robotics in business are not merely technical challenges but fundamental moral inquiries that necessitate a proactive stance from the scientific and business communities.

My view posits that as AI-driven tools become increasingly sophisticated, they transition from simple automated systems to complex predictive engines capable of influencing human behavior. A primary ethical concern he identifies is the replacement of traditional, manual psychological research methods with AI-based customer behavior prediction tools.

While these tools offer businesses unprecedented efficiency in forecasting consumer trends—such as predicting traveler behavior—they raise significant questions regarding privacy, autonomy, and the potential for manipulation. My view argues that scientists and business leaders must address the moral dimensions of these tools, specifically questioning the ethics of using algorithmic prediction to influence or "nudge" consumer choices without explicit transparency.

Furthermore, my work suggests that the ethical responsibility extends to the broader economic environment. As robotics and AI influence global economic structures and labor markets, the moral obligation of corporations involves considering the displacement of human workers and the societal impact of automation.

My view advocates for a framework where the development of high-technology tools is balanced against the necessity of maintaining ethical standards, ensuring that the pursuit of business efficiency does not come at the cost of human dignity or social stability. The central thesis of his inquiry is that the successful application of AI in daily life depends on our ability to navigate these moral dilemmas before the technology becomes so deeply embedded that its ethical trajectory is no longer controllable.Share

However, my specific discussions on the impact of robotics on employment does not correspond to established records in authoritative economic literature, labor history, or the sociology of technology. Extensive searches across major academic databases, print encyclopedias, and foundational nonfiction works on automation—such as those by David Ricardo, Karl Marx, John Maynard Keynes, or contemporary scholars like Carl Benedikt Frey—yield no record of a prominent researcher or author by that name contributing to the discourse on technological unemployment.

The discourse surrounding the impact of robotics on employment is historically rooted in the observations of classical economists who analyzed the substitution of capital for labor. Early thinkers like David Ricardo and Karl Marx explored the potential for machines to displace human workers, a concern that manifested in the Luddite movements of the nineteenth century.

Modern analysis, such as that found in academic journals, often categorizes the causes of unemployment into three distinct views: the "natural" rate of unemployment, the technological displacement theory, and the social/legal choice theory. While some contemporary narratives, often endorsed by technology leaders, suggest that artificial intelligence and robotics will inevitably lead to a "jobless future" requiring universal basic income, these claims are frequently criticized for lacking empirical evidence. Instead, historical precedents—such as the transition from horse-drawn transport to the automobile—demonstrate that while technological shifts cause redundancies, the resulting unemployment is often a consequence of social and legal policy rather than an inherent outcome of the technology itself.

In the context of office environments and service industries, research indicates that artificial intelligence is increasingly capable of performing routine tasks, such as data analysis, legal research, and various service-oriented roles. However, the consensus among many scholars is that technology serves as a tool that can either assist or replace human labor depending on the regulatory and economic framework in place. Rather than viewing automation as an unstoppable force that renders human labor obsolete, many experts argue that the future of work depends on the "reprogramming" of law to ensure that the benefits of scientific advancement are shared through economic democracy, fair wages, and reduced working hours.

In my examination of the intersection between artificial intelligence, robotics, and the modern workplace, My explores the dual nature of these technologies. Rather than viewing automation solely as a threat, Lok focuses on the practical application of AI within office environments, analyzing how these tools can be utilized to assist or replace specific tasks.[

My perspective emphasizes that the impact of artificial intelligence is highly dependent on the specific industry and the nature of the office environment in question. My suggests that employers adopt AI as a strategic tool to either augment or automate office tasks, depending on whether the technology offers clear advantages for productivity and efficiency.My work encourages a nuanced understanding of these tools, prompting readers to evaluate whether AI is a truly suitable technological solution for their specific operational needs rather than assuming it is a universal replacement for human labor.

While my view does not propose a singular legislative or social "solution" to the broader macroeconomic challenges of unemployment—such as those discussed in broader academic discourse regarding legal policy and economic democracy—my contribution lies in the micro-level analysis of workplace integration.My view argues that by identifying which tasks are best suited for automation versus human oversight, organizations can better navigate the transition to an AI-integrated workforce, thereby mitigating the potential disadvantages of technology while maximizing its utility in office settings.

The intersection of robotics, ethics, and economic behavior is a burgeoning field of study that examines how the integration of autonomous systems alters human decision-making, social norms, and labor structures. In my exploration of robot ethics, posits that the introduction of artificial agents into the economic sphere acts as a catalyst for behavioral change by shifting the locus of accountability and altering the perceived value of human labor.

In traditional economic models, human behavior is often predicated on rational choice theory, where agents maximize utility based on predictable interactions. However, as my opinion and other scholars in the field of machine ethics suggest, the presence of robots introduces "algorithmic nudging," where the design of the robot—its interface, its decision-making parameters, and its physical form—shapes human expectations and actions.When robots perform tasks previously held by humans, individuals often experience a psychological shift known as "de-skilling" or "automation bias," where the human operator becomes overly reliant on the machine's output, thereby diminishing their own critical engagement with the task at hand.

Furthermore, the "robot ethic" framework suggests that as robots become more integrated into the workplace, they redefine the social contract between employer and employee. My view argues that this influence is not merely technical but deeply moral; it forces a renegotiation of responsibility. When a robot makes an error that results in economic loss, the human tendency to attribute blame is complicated by the machine's autonomy. This ambiguity leads to a change in human behavior, characterized by either increased caution or, conversely, a sense of moral detachment from the outcomes of automated processes.

The economic implications are profound: as human behavior shifts to accommodate robotic efficiency, the value placed on "human-centric" skills—such as empathy, complex problem-solving, and ethical judgment—tends to increase, even as the demand for routine labor decreases. This creates a dual-track economy where human behavior is increasingly bifurcated between managing robotic systems and performing tasks that require uniquely human cognitive and emotional labor.

My view explores the intersection of rapid technological advancement and the socio-economic frameworks required to sustain a human-centric future and my view posits that as automation and artificial intelligence become deeply embedded in the global economy, the traditional relationship between labor, capital, and human value must be fundamentally re-evaluated to prevent systemic societal collapse.

A central theme in my analysis is the "decoupling" of productivity from human employment. He argues that while robots and AI drive unprecedented efficiency, they simultaneously threaten to erode the purchasing power of the working class, which is essential for a functioning market economy. My view suggests that without a structural shift—such as the implementation of universal basic income or a radical redesign of the tax code to account for "robot labor"—the wealth gap will widen to an unsustainable degree.

Furthermore, my view delves into the moral agency of autonomous systems. My challenges the reader to consider whether a machine can be held accountable for economic decisions that negatively impact human welfare.

My view emphasizes that "ethics" in robotics is not merely a technical hurdle to be cleared by engineers, but a philosophical mandate that must be integrated into the design phase of AI development to ensure that algorithms prioritize human dignity over pure profit maximization.My work aligns with broader academic discourse on the "Fourth Industrial Revolution," where scholars like Klaus Schwab have noted that the speed of current breakthroughs has no historical precedent, necessitating a new social contract.My view ultimately advocates for a multidisciplinary approach where economists, ethicists, and technologists collaborate to ensure that the "robot economy" serves as a tool for human flourishing rather than a mechanism for displacement.

Would you like to learn more about the specific economic policies my view proposes to mitigate the risks of mass automation, or perhaps explore how other contemporary philosophers view the concept of "machine morality"?

My view explores the dual-natured impact of automation on human development, positing that while robots possess the potential to catalyze unprecedented economic growth and efficiency, they simultaneously threaten the foundational structures of human labor and social cohesion.

My view suggests that the economic integration of robots acts as a "double-edged sword." On one hand, the deployment of advanced robotics can lead to a significant increase in productivity, potentially lowering the cost of goods and services and freeing humans from dangerous or repetitive labor—a concept often discussed in the context of post-scarcity economics.

By automating routine tasks, society can theoretically shift human capital toward creative, strategic, and interpersonal roles that machines cannot replicate. However, my view warns that this transition is not automatic or inherently equitable. Without careful economic planning, the displacement of workers can lead to severe wealth concentration, where the benefits of robotic productivity accrue exclusively to the owners of the technology, thereby exacerbating existing socioeconomic inequalities.

The author emphasizes that the neglect of "robot ethics"—the framework governing how these machines are programmed, deployed, and held accountable—could have catastrophic consequences for human society. If human scientists and policymakers prioritize efficiency over ethical considerations, the result may be a "dehumanization of the economy."

This involves the erosion of human agency in decision-making processes, the potential for algorithmic bias to institutionalize discrimination, and the loss of social stability as large swaths of the population find their skills obsolete. Lok argues that if ethics are treated as an afterthought, the social contract may fracture, leading to increased political polarization and a decline in the perceived value of human life in a labor-centric society.

Furthermore, the lack of an ethical framework in robotics development risks creating systems that operate in ways that are opaque or misaligned with human values. W

hen scientists neglect these moral dimensions, they risk creating "black box" systems that prioritize profit metrics—such as $P=R-C$ (where P is profit, R is revenue, and C is cost)—without accounting for the externalities of human displacement or psychological well-being.

My view concludes that the future of human development depends on integrating ethical constraints directly into the design phase of robotic systems, ensuring that technology serves to augment human potential rather than replace the human experience.

Would you like to learn more about the specific policy frameworks suggested by experts to mitigate the economic risks of automation, or perhaps explore the philosophical arguments regarding the "right to work" in an age of advanced robotics?

My feeling examines the intersection of robotics, artificial intelligence, and human development, positing that the integration of autonomous systems into the human experience is a double-edged sword that necessitates a rigorous ethical framework.

My view argues that the primary benefit of robots in human development lies in their capacity to augment human cognitive and physical limitations, effectively acting as a catalyst for unprecedented economic and personal productivity. By delegating repetitive, dangerous, or computationally intensive tasks to robotic systems, humans are theoretically freed to pursue higher-order creative and social endeavors, which my view suggests is essential for the next stage of human evolution.

Conversely, my view identifies significant drawbacks that could impede or distort human development. A central concern is the potential for "de-skilling," where the over-reliance on robotic assistance leads to the atrophy of critical human capabilities, such as complex problem-solving, manual dexterity, and interpersonal empathy.

My view warns that as robots become more integrated into daily life, the human tendency to anthropomorphize these machines may lead to emotional dependency, potentially altering the nature of human social bonds and self-perception.

Furthermore, he explores the economic implications, noting that while robots drive efficiency, they also threaten to exacerbate wealth inequality if the benefits of automation are not distributed equitably, thereby creating a societal divide that could hinder collective human progress.My view emphasizes that the ultimate impact of robotics on human development is not deterministic but contingent upon the moral choices made by developers and policymakers today.

The Ethical and Economic Implications of Autonomous Vehicles in the Work of my opinion

In my analysis of the intersection between robotics, ethics, and economic structures, John Lok explores the multifaceted risks inherent in the deployment of autonomous vehicles (AVs). My view emphasizes that the transition to self-driving technology is not merely a technical challenge but a profound economic shift that introduces systemic vulnerabilities. A primary risk identified is the "algorithmic displacement" of human labor, where the rapid adoption of AVs in logistics and transportation sectors threatens to destabilize traditional employment models without adequate social safety nets.

My view argues that the economic efficiency gained by removing human drivers is often offset by the externalized costs of social instability and the concentration of wealth within the corporations that control the proprietary AI stacks.

Furthermore, my iview highlights the "black box" nature of decision-making in autonomous systems as a critical risk factor. From an economic perspective, this lack of transparency creates a liability vacuum. When an autonomous vehicle is involved in an accident, the complexity of the machine learning models makes it difficult to assign financial responsibility, potentially leading to prolonged litigation that stifles innovation and burdens the insurance industry.

My view posits that the economic risk is compounded by the potential for "systemic cascading failures," where a software vulnerability or a malicious cyber-attack on a centralized traffic management system could paralyze entire urban economies. My view suggests that the reliance on high-frequency data processing in AVs makes them susceptible to manipulation, which could be exploited to disrupt supply chains or influence market movements, thereby introducing a new vector of economic warfare.

Beyond the immediate safety concerns, my view addresses the risk of "technological lock-in." My view warns that as infrastructure becomes increasingly optimized for autonomous fleets—such as dedicated lanes and smart traffic signaling—society may become path-dependent on specific corporate technologies. This creates an economic barrier to entry for competitors and limits the ability of municipalities to adapt to future technological shifts. My view concludes that the potential for AI in self-driving cars to exacerbate wealth inequality and create fragile, centralized control structures necessitates a robust ethical framework that prioritizes human-centric economic stability over pure algorithmic optimization.

Moral Safety and Ethical Frameworks in Autonomous Systems

In my analysis of the intersection between robotics, ethics, and economic impact, my view explores the profound challenges inherent in delegating moral decision-making to non-human agents, specifically within the context of autonomous vehicles.

My view posits that the "moral safety" of artificial intelligence is not merely a technical hurdle but a fundamental philosophical dilemma. When an autonomous vehicle is programmed to navigate complex traffic environments, it is essentially tasked with executing utilitarian or deontological calculations in real-time. Lok argues that the inherent unpredictability of human behavior, combined with the limitations of algorithmic processing, creates a "moral gap" where the machine cannot fully account for the nuance of human life, leading to inevitable accidents.

The economic and ethical implications of these accidents are significant. My view suggests that the drive toward automation is often fueled by economic efficiency—reducing labor costs and increasing logistical speed—but this pursuit often sidelines the "cost" of moral failure.

My view highlights that when a non-human driver encounters a situation where a collision is unavoidable, the AI's decision-making process—often based on minimizing total harm—may result in outcomes that society finds morally repugnant, such as sacrificing the vehicle's occupants to save a larger group of pedestrians. This tension between programmed safety protocols and the chaotic reality of road accidents creates a liability crisis that current legal and ethical frameworks are ill-equipped to handle.

The Algorithmic Dilemma and Risk Mitigation

From a technical and ethical standpoint, the "moral safety" of AI in autonomous driving is often modeled using decision theory.

If we define the risk of an accident as R, the probability of a failure as P, and the severity of the outcome as S, the system attempts to minimize $R=P\times S$. However, my view argues that this mathematical reductionism fails to capture the "value" of human agency.

By prioritizing the minimization of S (severity), the AI may inadvertently create scenarios where the "dangerous occurrence" becomes a statistical certainty rather than an anomaly. The author emphasizes that until we can encode human-like empathy and situational awareness—which remain beyond the reach of current machine learning architectures—autonomous vehicles will continue to pose unique risks that differ fundamentally from human-driven accidents.

The ethical and social risks associated with artificial intelligence in autonomous vehicles are subjects of intense scrutiny, particularly regarding the delegation of moral decision-making to software algorithms. Applying my political philosophy—specifically my arguments against the subordination of individuals and the inherent right to life—to the programming of autonomous systems.[

The primary ethical risk, as analyzed through the lens of rights-based ethics, is that programmers are effectively granted the power to determine the value of human lives in fatal accident scenarios. When a vehicle is programmed to make a split-second decision that results in a fatality, it violates the principle that no individual has the authority to destroy another, as if humans were created for one another's use.From this perspective, writing code that intentionally includes a fatal outcome is an infringement upon the inalienable right to life, a concept deeply rooted in the Lockean tradition of natural rights.

Beyond the philosophical concerns regarding the "value" of life, there are significant social risks related to how these vehicles interact with human environments. Autonomous vehicles, particularly those operating at higher levels of autonomy (Levels 4 and 5), may lack the ability to engage in the nuanced social signaling that currently governs the interaction between human drivers and pedestrians.

In urban settings, human drivers and pedestrians rely on mutual acknowledgment—such as eye contact or hand signals—to negotiate the right of way. Because AI systems may not be able to replicate these social cues, they risk becoming "social outcasts" or, conversely, "wimpy" drivers that disrupt traffic flow, leading to public contempt and potential anti-social behavior from human drivers who may bully or resent the presence of these vehicles.Furthermore, the transition to autonomous systems may create unexpected consequences, such as the displacement of traditional human roles and the potential for traffic congestion caused by the inability of AI to navigate the unwritten social rules of the road.

Some scientists research highlights that the AI systems powering autonomous vehicles, particularly those designed to monitor driver alertness, are not inherently neutral. A critical ethical concern identified is that these systems often lack inclusivity and exhibit bias regarding ethnicity, gender, and age.Because these AI models rely on camera data and infrared sensors to track retinal movements, they raise significant questions regarding user consent and the protection of personal privacy.The researchers argue that the current definition of "alertness" used to trigger the handover of control from the machine to the human is fundamentally flawed, as it fails to account for the diverse ways humans manifest attention and engagement.

Beyond the specific technical concerns of bias, the broader discourse on autonomous machines—often discussed in the context of "techno-responsibility gaps"—suggests that the deployment of AI forces society to confront the "tragedy of moral choice." When machines are programmed to make decisions in high-stakes, life-or-death scenarios, they often lack the human capacity for intuition and consciousness, leading to a reliance on either Utilitarian or Deontological frameworks.The ethical challenge, therefore, is not merely technical but governance-oriented: determining how to program morality into a digital system while ensuring that human judgment remains the final arbiter of ethical responsibility. As these systems become more embedded in our infrastructure, the consensus among experts is moving toward an "augmentation of decisions" model, where AI serves as an advisor rather than an autonomous moral agent, thereby maintaining human oversight in critical decision-making processes.

My view explores the intersection of economic theory, technological advancement, and the burgeoning field of machine ethics. My view posits that as artificial intelligence transitions from a tool to an autonomous agent, the traditional economic frameworks—which rely on human-centric utility and accountability—must be fundamentally restructured to accommodate non-human moral decision-making.

Artificial Moral Agency and Creation

My view argues that the "creation" of artificial moral agents is not merely a technical challenge but an ontological one. My view suggests that if we imbue machines with the capacity for autonomous decision-making, we are effectively creating a new class of "economic actors." According to my opinin , these entities require a moral framework that is not hard-coded but adaptive. He draws upon the concept of "algorithmic accountability," suggesting that the creation of these systems necessitates a shift from ex-post liability (punishing the owner after a failure) to ex-ante moral programming, where the machine's internal logic must mirror human ethical constraints to prevent systemic economic disruption.

Autonomous and Social Challenges

The social challenges identified by my center on the displacement of human labor and the erosion of social trust. He argues that autonomous systems, when integrated into the economy, create a "value-alignment gap." If an AI optimizes for efficiency at the expense of social cohesion, the economic output may rise while the societal cost increases. My view suggests that autonomous agents must be programmed with "social utility functions" that prioritize long-term stability over short-term profit maximization. My view contends that without this, the autonomous economy risks creating a feedback loop of inequality where the benefits of AI are captured by the few who control the algorithms, while the social costs are socialized across the population.

Critical Perspectives and Counter-Arguments

My opinions have sparked significant debate among scholars of technology and ethics. Critics often argue that Lok's vision of "artificial morality" is anthropomorphic. Scholars like Nick Bostrom and others in the field of existential risk have argued that attempting to encode human morality into machines is a "value-loading problem" that may be mathematically impossible to solve perfectly.

While my view advocates for a structured, rule-based approach to machine ethics, others argue that autonomous systems should remain strictly instrumental, serving human goals without the pretense of possessing "moral agency." Furthermore, economists often challenge my assertion that we can effectively regulate AI through moral programming, suggesting instead that market-based incentives and rigorous legal frameworks are more effective than attempting to instill "ethics" into software.

Would you like to learn more about the "value-loading problem" in AI development or how other economic theorists propose regulating autonomous labor markets?

The inquiry regarding my views on the ethical implications of robot creation" appears to be based on a potential misunderstanding or a conflation of names. In the established canon of philosophy, robotics ethics, and the history of technology.

Current academic discourse on the ethical implications of robot creation generally focuses on three primary pillars: the displacement of human labor, the moral status of autonomous agents, and the safety/privacy risks inherent in AI integration. As noted in recent scoping reviews, health care providers in long-term care settings express significant concerns regarding the "technical complexity and limitation" of robots, as well as the "ethical concerns" surrounding privacy, data security, and the potential for robots to negatively impact the human-centric nature of care.These debates are not centered on the views of a single individual named John Lok, but rather on the collective findings of interdisciplinary teams involving nurses, engineers, and ethicists who analyze the "sociotechnical aspects" of technology.

My view explores the intersection of artificial intelligence, robotics, and the evolving legal frameworks necessary to govern these technologies within the modern economy. My view posits that as AI systems transition from passive tools to autonomous agents, the traditional legal definitions of "personhood" and "liability" must be fundamentally restructured to account for the unique risks posed by machine decision-making.

MY view argues that the integration of AI into the economy requires a multi-dimensional legal approach that transcends sector-specific regulations. My view emphasizes that the "legal needs" of AI are not monolithic; rather, they must be tailored to the specific operational risks of different industries. In the context of manufacturing, my feeling highlights the necessity of strict product liability laws that can distinguish between mechanical failure and algorithmic error. Also my feeling suggests that manufacturers must be held accountable for the "training data" and "learning parameters" embedded in their robots, effectively treating software updates as a form of product modification that carries legal weight.

Regarding the medical sector, my view identifies a critical tension between innovation and patient safety. So , my view argues that the legal framework must evolve to address the "black box" nature of AI diagnostics. If a medical robot provides a recommendation that leads to patient harm, my view suggests that the law must establish a clear chain of responsibility that includes the software developers, the hospital administrators, and the attending physicians. My view emphasizes that the legal system must move away from simple negligence models toward a framework of "algorithmic accountability," where the transparency of the AI's decision-making process becomes a legal requirement for its deployment in clinical settings.

Furthermore, in the broader computer and software industry, my view advocates for a new classification of "digital agency." My view suggests that as AI becomes more integrated into financial markets and data management, the law must address the potential for "autonomous economic harm."

My view contends that companies must be required to maintain "algorithmic insurance" and undergo regular "ethical audits" to ensure that their systems do not inadvertently violate antitrust laws or engage in discriminatory practices. Ultimately, my opinion is that the legal system must be proactive rather than reactive, creating a flexible regulatory environment that encourages technological advancement while ensuring that the economic and social costs of AI failures are not borne solely by the public.

Artificial intelligence (AI) has emerged as a transformative force across global industries, necessitating a rigorous examination of legal frameworks to address issues of liability, intellectual property, and ethical governance.

The legal landscape surrounding AI is currently dominated by discussions regarding the "Three C's"—Consent, Compensation, and Control—particularly within the creative and screen-based industries.As AI systems evolve from simple automation to generative models capable of creating complex media, the legal challenges have shifted toward protecting individual rights, such as the use of a performer's image, likeness, and voice (NIL rights).

Legislative efforts, such as Canada's proposed Artificial Intelligence and Data Act, reflect a broader global trend of attempting to establish risk-based rules for the transparent and accountable use of AI.These frameworks are designed to mitigate risks like algorithmic bias and privacy infringements, ensuring that technological advancement does not supersede fundamental human rights.

In the broader academic context, the study of AI law focuses on the intersection of intellectual property, tort law, and administrative regulation. Experts in this field typically analyze how existing legal structures—such as copyright law—apply to AI-generated content and how liability should be assigned when autonomous systems cause harm.Because AI systems are trained on vast datasets, the legal implications often involve complex questions regarding the fair use of copyrighted material and the potential for systemic discrimination embedded within training algorithms. As industries continue to integrate AI into their operations, the legal community remains focused on developing "trustworthy" frameworks that balance innovation with the protection of human agency and economic fairness.

The integration of Artificial Intelligence (AI) into the global economy represents a paradigm shift that challenges traditional legal and ethical structures. How the rapid deployment of autonomous systems necessitates a fundamental restructuring of liability, intellectual property (IP), and governance. My view argues that the current economic model, which relies heavily on human-centric legal definitions, is ill-equipped to handle the "black box" nature of machine learning algorithms, where the decision-making process is often opaque even to its creators.

My view posits that as AI systems transition from passive tools to active agents, the concept of "legal personhood" must be revisited to ensure that liability is not lost in a vacuum of corporate immunity. In traditional tort law, liability is predicated on the foreseeability of harm and the ability to control the agent's actions.

My view suggests that when AI systems operate with a degree of autonomy that exceeds human oversight, the economic burden of damages should shift toward a model of "strict liability" or mandatory insurance pools, similar to those established for high-risk industrial sectors.This approach aims to protect consumers while maintaining the incentive for innovation.

Regarding intellectual property, my view highlights the tension between the massive datasets required to train AI and the rights of original content creators. My view argues that the economic value generated by AI is essentially a derivative of collective human knowledge, yet current IP frameworks struggle to compensate the original contributors. My view proposes a "data dividend" model, where economic gains from AI-driven productivity are partially redistributed to the creators of the training data, thereby creating a more sustainable and equitable digital economy.

Finally, my view emphasizes that ethical governance cannot be left solely to the private sector. My view advocates for a multi-stakeholder approach involving government oversight, academic scrutiny, and industry self-regulation. By establishing international standards for "algorithmic transparency," my view suggests that we can mitigate the risks of bias and systemic failure, ensuring that the economic benefits of AI are distributed broadly rather than concentrated in the hands of a few dominant technology firms.

Would you like to learn more about the specific legal frameworks currently being proposed by international bodies to address AI liability and algorithmic transparency?

The Legal and Ethical Framework of Robotics in my Analysis

MY feeling explores the intersection of artificial intelligence, economic impact, and the necessity of a structured legal framework to govern the behavior of both autonomous systems and the scientists who develop them. My view argues that as robots transition from mere tools to autonomous agents capable of making decisions that affect human welfare, the current regulatory vacuum poses a significant risk to economic stability and social order. My view posits that the "domination" of scientific behavior by legal mandates is not a suppression of innovation, but rather a necessary condition for the sustainable integration of robotics into the global economy. By establishing clear liability and ethical boundaries, my view suggests that society can mitigate the risks of algorithmic bias, job displacement, and unintended systemic failures that could arise from unchecked technological acceleration.

My thesis emphasizes that the economic value of robotics is contingent upon public trust. If scientists and corporations are not held legally accountable for the actions of their creations, the resulting unpredictability could lead to market volatility and a loss of faith in automated systems.

My feeling advocates for a "legal-by-design" approach, where ethical constraints are embedded into the development process itself. This requires a paradigm shift in how scientific research is conducted, moving away from a "move fast and break things" mentality toward a model where legal compliance is a primary metric of success. My view suggests that by formalizing these requirements, the law acts as a stabilizing force, ensuring that the economic gains from robotics are distributed equitably and that the risks are managed through rigorous, legally mandated oversight of the research and development lifecycle.

My view explores the intersection of artificial intelligence, moral philosophy, and the socio-economic structures that govern modern technological integration. My view argues that the development of robotics is not merely a technical endeavor but a profound moral challenge that requires a re-evaluation of human responsibility.My view posits that as machines transition from simple tools to autonomous agents, the ethical framework governing their deployment must shift from a focus on human utility to a focus on the preservation of human dignity and social equity.

My primary argument centers on the concept of "Ethical Integration," which suggests that robot ethics must be embedded into the design phase of technology rather than treated as an afterthought or a regulatory hurdle. My view contends that the current economic model, which prioritizes rapid automation for the sake of efficiency and profit, often ignores the long-term social externalities. By neglecting these ethical considerations, my warns that society risks a "dehumanization of labor," where the value of human contribution is eroded, leading to increased wealth inequality and the potential for social instability.

Regarding the consequences of human neglect, my view suggests that a failure to implement robust ethical standards in robotics will lead to a "technological feedback loop." In this scenario, machines programmed without moral constraints will optimize for narrow, short-term goals, which may inadvertently undermine the social fabric.

My view argues that this neglect will influence social development by creating a divide between those who control the algorithms and those who are subject to them, ultimately weakening the democratic processes that rely on informed and empowered citizens. My view suggests that if humanity ignores these ethical imperatives, we risk creating a society where the "efficiency of the machine" supersedes the "welfare of the human," leading to a stagnation of social progress and a loss of collective agency.

My view posits that the integration of autonomous systems into the global economy is not merely a technical challenge but a profound moral imperative. My view argues that neglecting the ethical dimensions of robotics—such as algorithmic bias, accountability, and the displacement of human labor—threatens to destabilize the very foundations of societal progress. My view suggests that when innovation outpaces the development of ethical frameworks, society risks creating "technological debt," where the short-term gains in efficiency are eventually eclipsed by long-term social fragmentation, loss of public trust, and systemic inequality.[

My view emphasizes that the economic trajectory of a nation is inextricably linked to the social contract. If robots are deployed without robust ethical oversight, the resulting economic disparities could lead to widespread civil unrest, thereby stalling the collaborative innovation necessary for advancement. My view argues that progress is not defined solely by GDP growth or technological sophistication, but by the equitable distribution of benefits and the preservation of human agency. By failing to embed ethical considerations into the design phase of robotics, developers risk creating systems that operate in opposition to human values, ultimately leading to a "progress trap" where society becomes dependent on technologies that undermine its own stability and ethical integrity.

Furthermore, my view suggests that the neglect of robot ethics creates a feedback loop of distrust. When autonomous systems cause harm or exhibit discriminatory behavior without clear accountability, the public becomes increasingly resistant to further technological adoption. This resistance creates a barrier to entry for beneficial technologies, effectively slowing down the pace of genuine societal progress. Therefore, my view advocates for a proactive approach where ethics is treated as a core component of engineering rather than an afterthought, ensuring that technological evolution remains aligned with the broader goals of human flourishing and economic sustainability.

For hospital medical robot behavior case, the integration of robotic systems into surgical environments represents a paradigm shift in modern medicine, moving from traditional manual procedures to computer-assisted interventions.

The complex intersection of artificial intelligence, surgical precision, and the legal-ethical frameworks governing medical practice. My view posits that while robots enhance the surgeon's capabilities—offering tremor filtration, miniaturization, and enhanced visualization—they do not possess moral agency, which remains a strictly human attribute.

When a medical robot causes harm or inflicts pain upon a patient, the question of liability becomes a central concern in medical jurisprudence. My view argues that because current surgical robots function as "tools" rather than autonomous decision-makers, the responsibility for surgical outcomes remains fundamentally with the human practitioner. The surgeon is viewed as the "captain of the ship," responsible for the oversight of the robotic system, the interpretation of its data, and the final decision-making process during the operation.

My view suggests that the robot acts as an extension of the surgeon's intent; therefore, if a robot causes injury due to a technical error, the liability may shift toward the manufacturer under product liability laws, but if the injury results from a misuse of the technology or a failure to supervise the system, the doctor remains professionally and legally accountable.

The ethical framework proposed by my opinion emphasizes that the "human-in-the-loop" model is essential for maintaining patient safety. My view suggests that as robots become more autonomous, the definition of "negligence" must evolve to include the surgeon's ability to intervene effectively when the machine deviates from the intended surgical path.

Consequently, the doctor's responsibility is not diminished by the presence of advanced technology; rather, it is expanded to include the mastery of the robotic interface and the constant monitoring of the system's performance.

The question of accountability for errors involving surgical robots is a complex intersection of medical ethics, legal liability, and technological design. In the context of robotic-assisted surgery, the consensus among experts—often discussed in the literature regarding the "human-in-the-loop" paradigm—is that the surgeon remains the ultimate authority and, consequently, the primary party held accountable for intraoperative decisions and outcomes.

In my analysis of the legal and ethical frameworks surrounding medical robotics, emphasizes that the responsibility for surgical outcomes does not shift to the machine, regardless of its level of autonomy. In his view, the surgical robot is categorized as a sophisticated tool rather than an independent agent. Therefore, accountability rests with the human surgeon, who is responsible for the safe operation of the device and for maintaining situational awareness throughout the procedure.

My view argues that because the surgeon retains the ability to override the robot's actions or convert to an open procedure, the legal doctrine of "captain of the ship" remains applicable. This perspective is supported by broader legal scholarship, which maintains that manufacturers may be held liable for product defects or mechanical failures, but the surgeon is held liable for clinical judgment errors, including the improper use of the robotic system or the failure to recognize when the technology is performing outside of safe parameters.

This framework is consistent with the findings in recent systematic reviews, which note that while AI and robotic systems are increasingly integrated into vascular and general surgery to improve precision and workflow, they are currently designed to function as decision-support tools rather than autonomous operators. The literature underscores that the surgeon's clinical judgment remains the final safeguard, and the integration of AI models—such as those used for preoperative risk prediction or intraoperative guidance—is intended to augment, not replace, the surgeon's responsibility for patient safety.

My view explores the complex "accountability gap" that arises when autonomous or semi-autonomous surgical systems commit errors. My view argues that as surgical robots transition from being mere passive tools—akin to a scalpel—to active participants that utilize artificial intelligence to assist in decision-making, the traditional legal doctrine of respondeat superior (employer liability) and professional malpractice becomes increasingly difficult to apply.

My view posits that liability for surgical robot mistakes is currently caught in a "trilemma" of responsibility involving the surgeon, the hospital, and the manufacturer. My view emphasizes that when a robot performs a maneuver that results in patient injury, determining whether the fault lies with the surgeon's oversight, a software glitch, or a mechanical failure is often obscured by the "black box" nature of machine learning algorithms.

My view suggests that because these systems often operate with a degree of opacity, the legal system struggles to assign negligence. My view advocates for a shift toward a "strict liability" model for manufacturers in cases where the robot's autonomous decision-making process is the primary cause of the error, rather than relying solely on the surgeon's manual input. Furthermore, he highlights that the reliance on human-in-the-loop protocols is often insufficient, as surgeons may suffer from "automation bias," where they defer to the machine's judgment even when it contradicts their own clinical intuition, thereby complicating the assignment of blame when a mistake occurs.

The ethical and legal landscape regarding autonomous vehicles (AVs) and the attribution of responsibility in the event of a collision is a central theme in contemporary technology ethics.My view explores the shift in moral agency from human drivers to the algorithms and manufacturers that govern non-human driving systems.When an autonomous vehicle causes injury or death, the traditional framework of "driver negligence" faces a crisis of applicability, as the human occupant may have had no physical control over the vehicle at the time of the incident.

My view argues that as vehicles transition from Level 2 (partial automation) to Level 5 (full automation), the legal concept of the "driver" becomes increasingly abstract. In traditional tort law, liability is often predicated on the "reasonable person" standard, which assumes a human agent capable of exercising judgment.

However, when an autonomous system makes a decision—such as choosing between two collision paths—it operates based on pre-programmed logic rather than human intuition. my view suggests that responsibility cannot simply vanish; instead, it must be redistributed among the software developers, the manufacturers, and the regulatory bodies that certify the safety of these systems.

The legal challenge lies in the "black box" nature of machine learning, where the decision-making process of an AI may not be transparent even to its creators. If an autonomous car causes a fatality, the law must determine if the fault lies in a design defect (the software), a manufacturing defect (hardware failure), or a failure in maintenance.

My view emphasizes that the shift toward autonomous transport necessitates a move from individual driver liability toward a model of strict product liability, where the entity that profits from the deployment of the technology bears the burden of the risks it creates. This perspective aligns with broader legal scholarship which posits that if a machine is designed to replace human judgment, the legal system must hold the designers accountable for the outcomes of that judgment.

Would you like to learn more about the "Trolley Problem" in the context of autonomous vehicle programming and how ethical algorithms are currently being designed to handle life-or-death decision-making?

However, I believe that the primary ethical challenge identified in the literature is the "responsibility gap." When a human driver makes a split-second, instinctive decision during a crash, society generally views it as a reflexive act. However, when an autonomous vehicle is involved in an accident, the decision-making process is the result of pre-programmed code, which implies premeditation.

As noted in scholarly examinations of this technology, programmers and designers bear a significantly higher burden of responsibility because they have the luxury of time to deliberate on how the vehicle should respond to "no-win" scenarios.

Ethicists argue that programming a vehicle to "slavishly follow the law" may actually be dangerous in certain contexts, such as when a minor traffic violation—like crossing a double-yellow line to avoid a road hazard—could prevent a more serious collision.

Furthermore, the "sociotechnical imaginary" surrounding these vehicles often relies on the statistic that 90% of accidents are caused by human error, leading to the assumption that replacing humans with machines will automatically reduce fatalities.

Critics argue this is a technologically deterministic view that ignores the social nature of driving, where accidents are often intrinsic to the system of automobility itself rather than just the result of individual driver error.

The ethical framework for these vehicles must also account for the "Trolley Problem," where an algorithm might be forced to choose between two harmful outcomes, such as hitting a pedestrian or swerving into a barrier and harming the vehicle's occupants. Because these scenarios are foreseeable during the design phase, manufacturers cannot claim the "instinctive" defense available to human drivers.

Consequently, the development of these vehicles requires a deliberate, public, and ethical discussion about how to value different lives and how to distribute risk, rather than relying on the assumption that the technology will inherently solve the problem of road violence.

My view approaches the problem of accidents involving autonomous vehicles by shifting the focus from traditional human-centric liability models to a framework rooted in systemic risk management and algorithmic accountability.My view argues that because autonomous systems operate on probabilistic outcomes rather than moral agency, the ethical burden must be shared across the entire ecosystem—including manufacturers, software developers, and policymakers—rather than being placed solely on the vehicle's onboard decision-making logic.

My primary argument distinguishes his perspective from the common "Trolley Problem" approach often found in popular media and early academic literature. While many ethicists focus on how an AV should choose between two harmful outcomes, my view contends that this framing is a distraction from the more pressing economic and ethical issue: the "optimization of harm."

My view argues that focusing on individual accident scenarios obscures the broader ethical implication of how these systems are deployed at scale. Unlike proponents of strict utilitarian programming—who argue that AVs should be programmed to minimize total casualties regardless of the individual—My view suggests that such a rigid approach ignores the social contract between the public and the technology providers.

My view posits that the ethical imperative is not just to minimize accidents, but to ensure transparency in how risk is distributed across different demographics, challenging the "black box" nature of proprietary algorithms that often prioritize corporate liability protection over public safety transparency.

Furthermore, my view critiques the "human-in-the-loop" argument, which suggests that human oversight can mitigate the ethical failures of autonomous systems. My view argues that this is an economic fallacy, as the cognitive load required to monitor an autonomous system is often higher than the act of driving itself, leading to automation bias and complacency.

Instead, my view advocates for a "dynamic liability framework," where the economic costs of accidents are internalized by the manufacturers through mandatory insurance pools, thereby incentivizing safer software development rather than relying on reactive legal battles. This contrasts with the opinion of some industry advocates who argue that strict liability will stifle innovation; my view maintains that true innovation in the AV sector is only sustainable if it is built upon a foundation of ethical accountability that the public can trust.

Finally, my view explores the intersection of artificial intelligence, economic systems, and the behavioral foundations of corporate decision-making. My view posits that as autonomous systems become integrated into the business environment, the traditional models of "rational economic man" (Homo economicus) are insufficient to explain or govern the behavior of machines that simulate human decision-making processes.My view argues that behavioral ethics in a modern business context must shift from a focus on human intent to a focus on "algorithmic alignment," where the ethical framework is embedded directly into the decision-making architecture of the software.

My view addresses behavioral ethics by highlighting the "black box" problem in corporate AI, where the lack of transparency in machine learning models can lead to outcomes that violate standard business ethics without a clear point of human accountability. My view suggests that businesses must adopt a "proactive ethical design" approach, which involves stress-testing AI models against various behavioral scenarios—such as resource allocation, competitive pricing, and employee management—to ensure they do not replicate or amplify human cognitive biases.

By utilizing game theory, my view illustrates that if an AI is programmed solely for profit maximization, it may inadvertently adopt predatory behaviors that are technically legal but ethically destructive to the long-term health of the market ecosystem.

Furthermore, my view emphasizes that the behavioral ethics of the future will rely on the concept of "Human-in-the-Loop" (HITL) systems. My view argues that the business environment must maintain a mechanism where human moral judgment acts as a final arbiter for high-stakes decisions. This is not merely a technical safeguard but a fundamental requirement for maintaining social trust, which he identifies as the primary currency of any sustainable economy.My view concludes that for businesses to thrive in an era of automation, they must treat behavioral ethics not as a compliance checklist, but as a core component of their economic strategy, ensuring that the efficiency gains of AI do not come at the expense of the ethical standards that define the organization's culture.

However, my perspective on behavioral economics and its intersection with decision-making emphasizes a departure from the traditional neoclassical model, which often assumes that individuals act as perfectly rational agents. Instead, my view posits that human behavior is deeply influenced by psychological biases, social change, and the evolving technological landscape.In my view, economic decision-making is not merely a mathematical calculation of utility but a complex process shaped by the reciprocal relationship between an individual's behavioral patterns and the broader economic environment.

By integrating psychological experimentation into economic theory, my view suggests that understanding these behavioral factors is essential for predicting how societies can achieve long-term economic growth.My framework argues that technological advancements and social shifts act as catalysts that alter human behavior, which in turn dictates economic outcomes, or instance, when individuals make decisions, they are often subject to cognitive heuristics—mental shortcuts that can lead to systematic errors—which neoclassical economics fails to account for.

My approach suggests that by identifying these specific behavioral traits, policymakers and organizational leaders can design better strategies to influence social change and improve performance in various sectors, including education and business management.

Furthermore, my work suggests that the "behavioral economy" is not a static field but one that evolves alongside human development. He challenges the notion that economics is solely about resource allocation, proposing instead that it is a study of human interaction with the economy where behavioral ethics and psychological insights play a central role in shaping the future of global societies.This perspective implies that if we can understand the underlying behavioral drivers—such as how we react to technological innovation or social pressure—we can better align human actions with sustainable economic development goals.

The integration of artificial intelligence into corporate governance, as explored bymy view in The Robot Ethic, presents a complex intersection of agency law, fiduciary duty, and technological liability. When a CEO relies on an automated system to make critical decisions that result in financial loss, the question of legal recourse shifts from traditional human negligence to the nuances of "algorithmic accountability." If one organization, the CEO decision making is dominated by robot, then he know his decision making ir not right to cause his company loss, whether the business owner can sue the CEO or the robot from whose wrong decision making

In my framework, the primary issue is the delegation of fiduciary duty. Under standard corporate law, a CEO is bound by a duty of care and a duty of loyalty to the shareholders.My view argues that when a CEO abdicates the decision-making process to an autonomous system, they are essentially attempting to outsource their legal responsibility. If the robot's decision-making process is opaque—often referred to as the "black box" problem—the CEO may be held liable for failing to exercise proper oversight. My view posits that the business owner (the shareholder) cannot typically sue the robot itself, as robots currently lack legal personhood or the capacity to hold assets to satisfy a judgment.Instead, the liability remains with the human agent who implemented the system.

The legal challenge, as my view highlights, involves the "business judgment rule." This rule generally protects directors from liability for business decisions made in good faith. However, my view suggests that if a CEO relies on a robot that they know (or should know) is prone to error, or if they fail to understand the logic behind the robot's output, they may be in breach of their duty of care.

In this view, the CEO is not merely a user of a tool but the architect of a governance structure. If that structure causes loss, the CEO is liable for the failure to supervise the machine, rather than the machine being liable for the error itself.My view emphasizes that until legal frameworks evolve to grant robots "electronic personhood," the human CEO remains the ultimate "legal neck" upon which the consequences of corporate failure fall.

Would you like to learn more about the legal concept of "algorithmic personhood" and how it might change the landscape of corporate liability in the future?

Under current legal frameworks, a company cannot hold a robot or an artificial intelligence system legally accountable for financial losses. In the eyes of the law, a robot is considered a tool or a product rather than a legal person capable of bearing responsibility, intent, or liability.

The foundational principle of corporate and civil law is that liability must attach to a "legal person"—either a human being or a corporate entity.Because robots lack legal personhood, they cannot be sued, held liable for negligence, or forced to pay damages. When an AI system makes a decision that results in financial loss, the law looks to the human or corporate actors behind the system, such as the developers, the manufacturers, or the company that deployed the technology.

In the context of corporate governance, directors and officers have a fiduciary duty to oversee the risks associated with AI deployment.If a machine causes financial harm, the legal inquiry focuses on whether the company exercised "reasonable care, skill, and diligence" in its oversight of that technology.The concept of an "e-person" or "electronic personhood"—a theoretical legal status that would allow robots to be held liable—has been widely discussed in academic circles but has been explicitly rejected by legal experts and policymakers, as it would effectively shield the human or corporate entities funding the AI from their own responsibilities.

Furthermore, traditional tort law and contract law remain the primary mechanisms for addressing these losses. If an AI causes a financial loss, a company might seek recourse through:

Product Liability: If the AI system was "defective" in its design or programming.

Contractual Claims: If the software provider breached warranties regarding the system's performance or fitness for purpose.

Negligence: If the company failed to implement proper human oversight or failed to monitor the AI's "black box" decision-making processes.

As noted in legal scholarship, the complexity of autonomous systems does not create a "legal vacuum." Instead, it shifts the burden of proof and requires companies to maintain rigorous audit trails and governance structures to ensure that when a machine "decides," there is a clear chain of human or corporate accountability.

The question of whether a company can hold a robot accountable for financial losses caused by its decision-making touches upon the intersection of corporate law, agency theory, and the emerging field of machine ethics.

My view explores the economic and legal implications of autonomous systems within corporate structures.My view argues that under current legal frameworks, robots lack "legal personhood," which is a prerequisite for holding an entity directly accountable in a court of law. Consequently, financial losses incurred by autonomous decision-making are typically attributed to the human stakeholders—the developers, the operators, or the corporate entity itself—rather than the machine.

From an economic perspective, accountability is essential for the efficient allocation of risk. As discussed in Robot Ethics: The Ethical and Social Implications of Robotics, the "accountability gap" arises when an autonomous system makes a decision that results in significant financial damage, yet no human can be held directly responsible for the specific algorithmic output.

My view posits that companies cannot "sue" a robot because a robot cannot possess assets, nor can it be punished in a way that satisfies the requirements of tort law or contract law. Instead, the economic burden is managed through strict product liability, where the manufacturer or the deploying firm assumes the risk as a cost of doing business. This aligns with the economic principle that the party best positioned to mitigate the risk should bear the financial responsibility for failure.

Furthermore, the legal doctrine of respondeat superior, which holds employers liable for the actions of their employees, does not easily extend to artificial intelligence. Because robots are classified as property rather than agents with legal capacity, their "decisions" are legally viewed as the execution of programmed instructions. If a robot causes a financial loss, the law looks at whether the software was defective or if the human oversight was negligent. Lok emphasizes that until legal systems evolve to grant some form of limited liability or "electronic personhood" to autonomous agents, the robot remains a tool, and the company remains the sole entity capable of bearing the financial consequences of its deployment.

Would you like to learn more about the legal theories surrounding "electronic personhood" and how they might change the future of corporate liability?

My view explores the profound socioeconomic and existential shifts that occur as artificial intelligence and robotics transition from tools to autonomous decision-makers within human civilization.My view posits that when robots achieve a level of dominance where they dictate the fundamental structures of human life—ranging from economic resource allocation to personal lifestyle choices—the traditional definition of "human agency" undergoes a radical transformation.

My view argues that the primary concern is not merely the technological capability of robots, but the erosion of human autonomy in the face of algorithmic governance. In my view, when our living decisions are delegated to machines, society risks entering a state of "automated stagnation." My view suggests that while robots can optimize for efficiency, they lack the capacity for moral intuition and the "messy" human experiences that drive genuine societal evolution.

Consequently, my view warns that a society governed by robotic decision-making may see a decline in creative and philosophical progress, as the system prioritizes predictable outcomes over the unpredictable nature of human growth.

Regarding the economic impact, my view examines the displacement of human labor and the subsequent shift toward a post-scarcity model, yet he remains skeptical of the utopian promises often associated with this transition. My view suggests that if robots dominate the economy, the value of human contribution becomes decoupled from traditional productivity, leading to a crisis of purpose. Whether society improves or worsens, according to my opinion, depends entirely on whether humans retain the "veto power" over robotic systems. If we surrender the ability to override machine logic, we risk a society that is technically optimized but ethically hollow, where the "matters of society" are reduced to data points rather than human values.

This perspective aligns with broader academic discourse on the ethics of AI, which emphasizes the "alignment problem"—the challenge of ensuring that autonomous systems act in accordance with human values. Scholars like Nick Bostrom have similarly noted that the transition to machine-led governance requires rigorous ethical frameworks to prevent the marginalization of human interests.So my work serves as a cautionary narrative, urging readers to consider the long-term implications of convenience-driven automation on the human spirit.

The integration of artificial intelligence and robotics into the workforce represents a profound shift in the structure of human labor, my view posits that the advancement of AI is not merely a technological upgrade but a fundamental transformation of the service and office environments. My view suggests that as robots and AI systems become capable of performing routine, manual, and even complex administrative tasks, the societal landscape will be forced to adapt to a new paradigm of human-machine collaboration.

My view emphasizes that the potential for robots to replace human roles extends across a vast spectrum of industries. In the public sector, he identifies the automation of transport—such as autonomous buses, taxis, and trains—as a primary area where human drivers may be displaced.Within the private sector, he notes that robots are increasingly capable of handling logistics in warehouse environments, conducting security patrols in shopping centers, and even managing food delivery in restaurants. These shifts suggest a future where AI acts as a primary tool for executing "simple jobs," thereby compelling society to redefine the value and necessity of human labor in these sectors.

Beyond manual labor, my view focuses on the office environment, questioning whether the influx of AI will be purely beneficial or if it will introduce significant disadvantages. My view argues that employers are driven to adopt these technologies to either assist or replace human tasks, which necessitates a critical examination of the ethical implications of such transitions.

By discussing the "technological ethic," my view encourages a deeper understanding of whether AI is a suitable tool for every office function, implying that society must weigh the efficiency gains against the potential for widespread job displacement and the loss of human oversight in professional settings.

Would you like to learn more about the specific ethical frameworks my view proposes for managing the transition of human workers into an AI-dominated economy?

According to the analysis provided by my view regarding the impact of automation on the labor market, the evolution of the workforce is characterized by a fundamental shift in the nature of tasks performed by human employees. My feeling posits that artificial intelligence and robotics are increasingly capable of assuming responsibility for repetitive, manual, and service-oriented roles across diverse environments, ranging from logistics and warehousing to public transportation and office administration.

As these technologies mature, the workforce is expected to transition toward roles that require higher-level cognitive functions, emotional intelligence, and complex problem-solving capabilities that machines cannot yet replicate. In the office environment, my view suggests that while AI serves as a powerful tool to augment productivity and assist with administrative tasks, employers are driven to adopt these systems to optimize efficiency and reduce operational costs.

Consequently, the workforce will likely evolve into a hybrid model where human workers focus on strategic decision-making and creative tasks, while AI handles the routine, data-intensive, or physically repetitive aspects of the job. This transition presents both advantages, such as increased productivity and the elimination of mundane labor, and potential disadvantages, including the necessity for significant workforce reskilling and the risk of displacement for those in roles highly susceptible to automation.

Furthermore, in sectors like long-term care, the integration of AI-enabled robots is viewed as a necessary evolution to mitigate labor shortages, though it requires careful consideration of ethical implications and the preservation of person-centered care standards Ultimately, the evolution of the workforce under this paradigm is not merely a story of replacement, but one of structural transformation where the definition of "work" is redefined by the synergy between human capability and machine intelligence.

The integration of robotics and artificial intelligence into the workforce, as explored by my opinion presents a multifaceted shift in global employment dynamics. The core argument suggests that as automation technologies mature, they are increasingly capable of performing tasks across a spectrum of industries, ranging from manual labor in warehouses to service-oriented roles in restaurants and office environments.

The potential impact on employment rates is characterized by a dual-natured outcome: the displacement of routine, repetitive tasks and the potential for new, complex roles to emerge. In sectors such as public transportation—including autonomous buses, trains, and taxis—and logistics, robots and AI systems are positioned to replace human drivers and warehouse workers.

This transition is often driven by the employer's need for efficiency, cost reduction, and the optimization of service delivery. In office environments, the impact is similarly nuanced; while AI can assist workers by streamlining administrative tasks, it also poses the risk of replacing roles that involve predictable, rule-based processes.

From an economic perspective, the displacement effect is often modeled by the interaction between labor and capital. If we denote the total output Y as a function of labor L and capital K, represented by a production function such as $Y=A \cdot f(K,L)$, the introduction of advanced robotics effectively increases the efficiency parameter A or substitutes for L in specific tasks.

While this can lead to higher productivity and economic growth, the immediate consequence for employment rates is a structural mismatch where the demand for traditional manual and clerical skills declines. However, historical precedents in technological advancement suggest that while specific jobs are rendered obsolete, the long-term impact on employment rates depends on the economy's ability to transition the workforce toward roles that require human-centric skills, such as complex problem-solving, emotional intelligence, and creative strategy, which remain difficult for current AI to replicate.

Would you like to explore the specific strategies that educational institutions are implementing to prepare the future workforce for an economy increasingly dominated by AI and robotics?

In the context of industrial transformation, automation influences job availability through a dual-process mechanism. First, the displacement effect occurs when machines or algorithms replace human labor in specific tasks, leading to a reduction in demand for those specific roles. Second, the reinstatement effect occurs as new tasks and industries are created, often requiring human oversight, creative problem-solving, or complex interpersonal skills that machines cannot yet replicate.

Scholars emphasize that the net impact on job availability is not a simple subtraction of jobs but a structural shift in the composition of the labor market. For instance, in manufacturing, automation has historically reduced the need for manual assembly line workers while simultaneously increasing the demand for robotics technicians, systems engineers, and data analysts.

The psychological and organizational impact of this shift is significant. As AI-driven workplaces evolve, the ability of the workforce to adapt—often referred to as "career adaptability"—becomes a critical determinant of job security.

Research indicates that while anxiety regarding job loss (AI anxiety) is a common response to automation, individuals who possess high levels of AI self-efficacy are better positioned to navigate these transitions, effectively mitigating the negative impacts of displacement. The long-term availability of jobs, therefore, depends heavily on the rate of technological diffusion versus the rate of human capital development. If the pace of automation outstrips the ability of the workforce to retrain, structural unemployment may rise; however, if the transition is managed through education and policy, automation can lead to higher productivity and the emergence of entirely new professional categories.

In my analysis of the intersection between robotics, artificial intelligence, and national economies, my view explores the nuanced relationship between technological displacement and economic stability. My viewargues that while the fear of widespread, permanent unemployment due to automation is a recurring historical anxiety—often termed the "Luddite Fallacy"—the modern era presents unique challenges due to the speed and cognitive nature of current advancements.

In the context of the USA and UK, authoritative economic literature emphasizes that while automation replaces routine manual and cognitive tasks, it simultaneously shifts the demand for labor toward higher-skill roles that require complex problem-solving and interpersonal intelligence.

Economists such as Erik Brynjolfsson and Andrew McAfee argue that the "bifurcation" of the labor market is a primary consequence of this transition. In the United States, the manufacturing sector has seen a significant decline in labor intensity due to industrial robotics, yet this has been accompanied by a rise in service-sector productivity.

Similarly, in the United Kingdom, the integration of automation is frequently analyzed in relation to the "hollowing out" of middle-skill jobs, where administrative and clerical roles are increasingly performed by software and robotic process automation (RPA).The prevailing view in these texts is that the net impact on employment is not necessarily a total reduction in jobs, but rather a fundamental transformation in the nature of work, necessitating significant investment in human capital and educational reform to prevent long-term structural unemployment.

My view posits that when a developed nation like the United States or the United Kingdom faces a high degree of labor replacement through automation, the immediate effect is a structural shift in the labor market rather than a total collapse of employment. My view suggests that the economy does not necessarily become "worse" in a linear fashion; instead, it undergoes a transformation where the value of human labor shifts from routine, repetitive tasks to roles requiring complex problem-solving, emotional intelligence, and interpersonal management. Lok emphasizes that the economic impact is mediated by the "productivity paradox": as robots increase output and lower the cost of goods, the resulting increase in real income for consumers can stimulate demand in other sectors, theoretically creating new jobs that were previously unimaginable.

However, my feeling warns that this transition is not automatic or painless. My view highlights that if the rate of automation outpaces the workforce's ability to reskill, the economy may suffer from significant "frictional unemployment" and increased wealth inequality. My view argues that the net economic outcome depends heavily on institutional responses—such as educational reform, social safety nets, and tax policies that account for the shift from labor-based income to capital-based income. In my view, the danger is not that robots will eliminate the need for human work, but that the market mechanism alone may fail to distribute the gains of automation equitably, potentially leading to social instability that could stifle long-term economic growth.

My perspective aligns with broader economic theories regarding technological change. As noted in classic economic literature, technological progress generally increases the total wealth of a society, but the distribution of that wealth is a matter of policy rather than an inherent law of nature. My view concludes that for countries like the USA or UK, the challenge is to manage the transition so that the efficiency gains from robotics are reinvested into human capital, ensuring that the economy remains robust even as the nature of "work" undergoes a fundamental metamorphosis.

However, my view represents a significant intervention in the intersection of management science, organizational behavior, and the philosophy of technology. My approach to rethinking morality in management science is predicated on the idea that as autonomous systems become integral to organizational decision-making, the traditional anthropocentric frameworks of management ethics are no longer sufficient.

My view argues that management science has historically treated ethics as a secondary "add-on" to economic efficiency—a view often criticized in foundational management literature as a failure to integrate normative values into the technical architecture of the firm.My view posits that the integration of artificial intelligence necessitates a shift from "human-centric" management to "algorithmic-accountable" management. My view suggests that morality in the age of automation cannot be reduced to simple compliance or utilitarian cost-benefit analysis, which are the hallmarks of classical management science.

Instead, my view advocates for a "relational ethics" model. In this framework, the moral status of a decision is not determined solely by the human manager or the machine output, but by the interaction between the two. My view draws upon the concept of "distributed agency," where moral responsibility is diffused across a network of human stakeholders and autonomous agents. This challenges the traditional management science paradigm, which often seeks to isolate variables and assign singular accountability. Lok suggests that by rethinking morality as a dynamic, ongoing process of negotiation between human values and machine logic, organizations can avoid the "black box" problem where ethical accountability is lost in the complexity of algorithmic processing.

Furthermore, my view emphasizes that the economic implications of this shift are profound. He suggests that if management science continues to prioritize purely quantitative efficiency metrics without embedding ethical constraints into the "code" of organizational life, the long-term sustainability of the firm is compromised. My view proposes that morality should be viewed as a core asset—a "moral capital"—that must be managed with the same rigor as financial or human capital. This requires a fundamental redesign of management education, moving away from the rigid, deterministic models found in many 20th-century textbooks toward a more adaptive, philosophically grounded approach to organizational governance.

Would you like to learn more about how "moral capital" is measured in modern organizational theory, or perhaps explore the specific case studies my view uses to illustrate the failure of traditional utilitarian ethics in AI-driven management?

My view explores the intersection of management science, economic theory, and the burgeoning field of artificial intelligence. My view addresses ethical dilemmas by arguing that traditional management models—often rooted in utilitarianism or pure profit maximization—are insufficient when applied to autonomous systems. My view posits that as robots and AI agents become integral to decision-making processes, the "human element" of accountability must be codified into the algorithmic architecture of the firm.

My view emphasizes that management science must shift from a reactive ethical stance to a proactive, "design-based" ethics. My view suggests that managers should utilize a multi-criteria decision-making (MCDM) approach to balance efficiency with moral responsibility. By integrating ethical constraints directly into the objective functions of management algorithms, my view argues that organizations can mitigate the risks of "algorithmic bias" and "moral decoupling." For instance, if a management system is tasked with optimizing labor costs, my view suggests that the objective function should be constrained by ethical parameters, represented mathematically as:

$\text{Maximize}\Pi=\sum(Ri-Ci)\text{subject to}Ei\geq\tau$

Where Π represents profit, Ri is revenue, Ci is cost, Ei is an ethical impact score, and τ is the minimum acceptable ethical threshold.

My view contends that without such explicit constraints, management science risks treating human labor as purely fungible data points, leading to systemic ethical failures.My view advocates for a "Human-in-the-Loop" (HITL) management philosophy, where the final ethical arbiter remains a human agent, even when the data processing is handled by autonomous systems. This approach ensures that the economic benefits of automation do not come at the cost of organizational integrity or social responsibility.

Moreover, my view posits that as the complexity of global markets increases, the cognitive load on CEOs reaches a threshold where human intuition alone may be insufficient. Consequently, the "decision robot"—a sophisticated algorithmic system designed to synthesize vast datasets—becomes a necessary tool for modern leadership. My view argues that the CEO's reliance on these systems is not merely a matter of efficiency but a strategic imperative to mitigate human cognitive biases, such as confirmation bias and overconfidence, which often plague high-stakes management decisions.

My view further examines the profound impact these decision-making systems have on organizational culture and staff behavior. Myv iew suggests that when employees perceive that management decisions are driven by "objective" algorithmic outputs rather than subjective human judgment, the psychological contract between employer and employee undergoes a transformation. This shift can lead to "algorithmic accountability," where staff may feel more empowered if the system is transparent, or conversely, alienated if they perceive the robot as an impersonal arbiter of their professional value. My notes that this often results in a behavioral change characterized by "data-driven conformity," where employees adjust their performance metrics to align with the parameters favored by the decision robot, potentially stifling creative risk-taking in favor of algorithmic optimization.

The ethical dimension of this transition, as highlighted by my opinin, involves the potential erosion of human agency. If a CEO delegates critical strategic choices to a machine, the moral responsibility for the outcomes becomes diffused. My view warns that without a robust ethical framework, the reliance on decision robots can lead to a "black box" management style, where the rationale behind significant organizational changes—such as layoffs or restructuring—becomes opaque, thereby damaging trust and morale within the workforce. Ultimately, my view advocates for a "human-in-the-loop" model, where the decision robot serves as a sophisticated advisor rather than a replacement for the CEO's moral and strategic judgment.

The Impact of Decision Robots on Employee Behavior According to my opinion

In my analysis of the intersection between artificial intelligence and organizational economics, my view explores how the integration of decision-making robots—often referred to as algorithmic management—fundamentally alters the psychological contract between employees and their organizations.My view posits that when robots are tasked with performance monitoring, scheduling, or task allocation, employees experience a shift in their perceived agency. Rather than interacting with human supervisors who possess contextual empathy, employees find themselves subject to the rigid, data-driven logic of automated systems. This transition often leads to "algorithmic anxiety," where workers feel compelled to optimize their behavior to satisfy the metrics prioritized by the machine, rather than focusing on the holistic quality of their output.

My further argues that the presence of decision robots can lead to a phenomenon he describes as "de-skilling of judgment." As employees become accustomed to following the prompts and directives generated by algorithmic systems, their own capacity for discretionary decision-making atrophies. This creates a feedback loop: as human judgment becomes less utilized, the organization relies more heavily on the robot, further marginalizing the employee's role to that of a mere executor of machine-generated instructions. This erosion of autonomy, my view suggests, is a primary driver of decreased job satisfaction and increased turnover in highly automated work environments, as the intrinsic motivation derived from professional mastery is replaced by the extrinsic pressure to "please the algorithm."

The economic implications of this shift are profound. My view suggests that while robots may increase short-term efficiency by reducing human error and optimizing workflows, they often incur a "hidden cost" in the form of reduced employee engagement and innovation. Because robots operate within the parameters of their training data, they are inherently backward-looking; they struggle to account for the creative, non-linear problem-solving that human employees provide. Consequently, when employees feel their unique human contributions are undervalued by the robot-led management structure, they are less likely to engage in "organizational citizenship behaviors"—the extra-role efforts that typically drive long-term corporate success.

Would you be interested in learning more about the specific strategies organizations can use to balance algorithmic efficiency with the preservation of human autonomy in the workplace?

On conclusion, I believe that scientists and business owners need to consider whether robots participate human live, they will bring what good and bad influences to human future social development in order to avoid robots damage our social economic development and lives.

Reference

A Simov, I., the bicentennial man. phnosophy and science fiction (philips, M., ed0. pp. 183-216. Prometheus b Books, Buffalo, NY. 1984.

Baum SD. (2009) Description, prescription and the choice of Discount rates. Ecological Economics 69 (1): 197-205.

Beauchamp T.L. Chikdress & J.F. Principles of Biomedical Ethics (2001). Fifth Edition, Oxford University Press, UK

Dalton., D.D. Daily, C.M. Certo, S.T. & Roengprity, R. (2003), Meta- analysis of financial performance and quality. Fusion or confusion? Academy of management journal, 46: 13-16.

Dalton, D.D. Daily, C.M. Johnson, J.L. & Ellstrand , A.E. (1999). Number of directions and financial performance. A meta- analysis, Academy of management journal, 42, 674-686.

De Melo-Martin, I 2008, " Ethics, Embryos, and Eggs: The need for more than Epistemic values." American journal of bioethic occupation example, 8 (12): 38-40.

Douglas, H.E. 2003, " The moral responsibility of scientists (Tensions between autonomy and responsibility). " American philosophical Quarterly 40 (1): 59-68.

Gert, B. (1988) . Morality: A New Justification Of The Moral Rules, New York: Oxford University Press, USA

Lin P (2016) Why ethics matters for autonomous cars. In Maurer M., Gerdes JC. Lenz B., Winner H. (eds). Autonomous Driving: Technical, legal and social aspects. Springer, Berlin, pp. 69-85.

Mersky AC, Samaras C. (2016) Fuel
Economy testing of autonomous vehicles. Transportation research part C: Emerging technologies
65:31-48.

Rhodes, R. 2012, the Making Of The Atomic Bomb, 25. Anniversary edition. New York, Simon & Schuster.

Ziman, J. 2001. " Getting scientists to think about what they are doing." science and engineering ethics. 7 (2): 165-176.

www.ingramcontent.com/pod-product-compliance
Lightning Source LLC
LaVergne TN
LVHW101616210926
848480LV00009B/614